Contemporary's

Level M

NUMBER POWER

Intermediate 1

McGraw Hill | Wright Group

Wright Group

ISBN 0-8092-0610-2

Copyright © 1999 by NTC/Contemporary Publishing.

Send all inquiries to:
Wright Group/McGraw-Hill
130 E. Randolph, Suite 400
Chicago, IL 60681

Manufactured in the United States of America.

11 12 RRO 10 09

Contents

To the Learner

Even if math has never been easy for you, this text will give you the instruction and practice you need to understand the basics. In this global and technological society, an understanding of math is important, and, at some time or another, you will be asked to demonstrate that you can solve math problems well.

Using *Number Power Intermediate 1, Level M* is a good way to develop and improve your mathematical skills. It is a comprehensive text for mathematical instruction and practice. Beginning with basic comprehension skills in addition, subtraction, multiplication, and division, *Number Power Intermediate 1, Level M* includes basic concepts about performing computations. For instance, concepts include information such as words like *plus, sum,* and *total* often signal addition problems; the addition symbol is a "+" sign, and addition problems have distinctive characteristics. Then problems, which illustrate the basic kinds of addition problems learners will confront, are included for plenty of practice.

Accompanying all of this practice are a Skills Inventory Pre-Test and a Skills Inventory Post-Test. The Skill Inventory Pre-Test will help you identify your math strengths and weaknesses before you begin working in the book. Then you can work in those areas where additional instruction and practice are needed. Upon completion of these exercises, you should take the Skills Inventory Post-Test to see if you have achieved mastery. Mastery is whatever score you and your instructor have agreed upon to be correct to insure that you understand each group of problems.

Usually mastery is completing about 80 percent of the problems correctly. After achieving mastery, you should then move on to the next sectionof instructionand practice. In this way the text offers you the chance to learn at your own pace, covering only the material that you need to learn. In addition, the instruction on each page offers you the opportunity to work on your own.

Addition, subtraction, multiplication, and division instruction and practice are presented in the first part of *Number Power Intermediate 1, Level M*. The other part contains applications. It includes numeration, number theory, data interpretation, pre-algebra, measurement, geometry, and estimation. You will work with ordinal numbers, place value of numbers, graphs, tables, charts, number sentences, calendars, time, plane figures, logical reasoning, problem solving, estimation, and many other topics.

Completing *Number Power Intermediate 1, Level M* will make you more confident about doing mathematical problems. Remember to use the Answer Key in the back of the book to check your responses. Soon you will find yourself either enjoying math for the first time or liking it even more than you did previously.

Skills Inventory Pre-Test

Part A: Computation

Circle the letter for the correct answer to each problem.

1 $10 \times 8 =$ _____

    A 80
    B 18
    C 88
    D 108
    E None of these

2
$$\begin{array}{r} 3,016 \\ + \ 2,963 \\ \hline \end{array}$$

    F 5,978
    G 6,979
    H 5,989
    J 5,979
    K None of these

3
$$\begin{array}{r} 114 \\ - \ 95 \\ \hline \end{array}$$

    A 39
    B 18
    C 19
    D 21
    E None of these

4
$$\begin{array}{r} 67 \\ \times \ 12 \\ \hline \end{array}$$

    F 704
    G 794
    H 201
    J 804
    K None of these

5 $7 \div 5 =$ _____

    A 2
    B 1
    C 1 r 2
    D 2 r 1
    E None of these

6 $72 \div 3 =$ _____

    F 14
    G 20 r 2
    H 24
    J 23
    K None of these

7 $446 + 282 =$ _____

    A 628
    B 728
    C 718
    D 638
    E None of these

8 $\dfrac{12}{15} - \dfrac{3}{15} =$ _____

    F $\dfrac{3}{5}$
    G 1
    H 9
    J $2\dfrac{2}{3}$
    K None of these

9 $6 \times \dfrac{4}{6} =$ _____

    A 4
    B $\dfrac{2}{3}$
    C $\dfrac{1}{3}$
    D $6\dfrac{2}{3}$
    E None of these

10 $34 \times 0.5 =$ _____

    F 120
    G 12.0
    H 1.2
    J 0.12
    K None of these

11 $\dfrac{2}{3} \times \dfrac{3}{4} =$ _____

    A $\dfrac{5}{12}$ **C** $\dfrac{5}{9}$
    B $\dfrac{1}{2}$ **D** $\dfrac{1}{6}$
    E None of these

12 $100 \times 0.435 =$ _____

    F 43.5
    G 435
    H 0.00435
    J 4.35
    K None of these

13 $1250 - 129 =$ _____

    A 1,111
    B 1,139
    C 1,121
    D 1,131
    E None of these

Part B: Applied Mathematics

Circle the letter for the correct answer to each problem.

14 What sign goes in the box to make the number sentence true?

12 ☐ 12 = 1

F ×
G ÷
H –
J +

15 What number is missing from this number sequence?

100, 80, _____, 40, 20

A 5
B 25
C 60
D 40

16 What time will the clock show in 20 minutes?

F 3:20
G 3:25
H 3:45
J 3:35

Keiko has weighed her baby every month for the last six months. This table shows her records. Study the table. Then do Numbers 17 through 19.

Age	Weight
2 days	6 pounds, 1 ounce
1 month	6 pounds, 9 ounces
2 months	7 pounds, 1 ounce
3 months	7 pounds, 8 ounces
4 months	8 pounds, 1 ounce
5 months	8 pounds, 9 ounces

17 How much weight did the baby gain during this time?

A 8 pounds, 9 ounces
B 2 pounds, 11 ounces
C 2 pounds, 8 ounces
D 2 pounds, 9 ounces

18 The baby has gained weight at a pretty steady rate. About how much did the baby gain each month?

F 8 ounces
G 1 pound
H 2 ounces
J 19 ounces

19 How much did the baby weigh at three months of age?

A $7\frac{1}{3}$ pounds

B $7\frac{1}{4}$ pounds

C $7\frac{1}{2}$ pounds

D $7\frac{4}{5}$ pounds

20 In 17,892, what does the 7 mean?

F 70 **H** 7,000
G 700 **J** 70,000

21 If the same number is used in both boxes, which of these statements would be true?

A If 7 − ☐ = 3,
then 3 − ☐ = 7.
B If 13 − ☐ = 8,
then 13 + 8 = ☐.
C If 7 + ☐ = 12,
then 12 ÷ ☐ = 7.
D If 3 × ☐ = 15,
then 15 ÷ ☐ = 3.

22 Lana is making ice for her party. She can only freeze 24 cubes at a time. If she makes 4 batches of ice cubes, how many cubes will she have?

F 28 **H** 86
G 64 **J** 96

23 Alli spent $26.78 for a shirt, $41.15 for some pants, and $72.59 for a pair of shoes. Which number sentence should he use to estimate to the nearest dollar how much he spent?

A $26.00 + $41.00 + $72.00 = ☐
B $27.00 + $41.00 + $73.00 = ☐
C $26.00 + $41.00 + $73.00 = ☐
D $27.00 + $41.00 + $72.00 = ☐

24 Which of these groups of numbers shows counting by fives?

F 145, 150, 155, 160, 165, 170
G 125, 150, 175, 200, 225, 250
H 55, 65, 75, 85, 95, 105, 115
J 50, 100, 150, 200, 250, 300

25 Don had $22.00 in his wallet. He withdrew another $40.00 from an ATM machine. Then he spent $12.00 for lunch. How much cash is left in his wallet?

A $60.00 **C** $28.00
B $10.00 **D** $50.00

This graph appeared in a newspaper article on teenage pregnancy. Study the graph. Then use it to do Numbers 26 through 28.

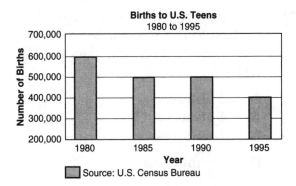

26 In which two years did about the same number of U.S. teens give birth?

F 1980 and 1985
G 1985 and 1990
H 1990 and 1995
J No two years had the same number.

27 During this period, about what was the average number of teens giving birth?

A 600,000
B 500,000
C 400,000
D 450,000

28 If the trend shown on this graph continues, about how many U.S. teens will give birth in the year 2000?

F 600,000 or more
G 500,000
H 450,000
J 400,000 or fewer

This table shows travel times as a commuter train makes four stops after our station. Use the table to do Numbers 29 through 31.

Travel Times from Central Station

Stop	Travel Time
Grant Street	12 minutes
Hohoken Street	16 minutes
Bentley Avenue	22 minutes
City Zoo	31 minutes

29 You want to travel from Central Station to the zoo and back. How much time would you spend on the train?

  A 31 minutes
  B 44 minutes
  C 62 minutes
  D $16\frac{1}{2}$ minutes

30 The travel time to Bentley Avenue is about $\frac{1}{3}$ of an hour. That is between which two amounts of time?

  F $\frac{1}{4}$ and $\frac{1}{2}$ of an hour
  G $\frac{1}{2}$ and $\frac{3}{4}$ of an hour
  H $\frac{3}{4}$ and $\frac{2}{3}$ of an hour
  J $\frac{1}{6}$ and $\frac{1}{5}$ of an hour

31 Which of these is the best estimate of the travel time from Bentley Avenue to the City Zoo?

  A 5 minutes
  B 10 minutes
  C 15 minutes
  D 20 minutes

32 What is the temperature shown on this thermometer?

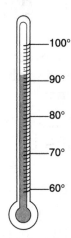

  F 90°F   **H** 92°F
  G 91°F   **J** 94°F

33 Hector must pay $102.50 in taxes. Which of the following amounts should he put on his check to the government?

  A one hundred twenty and—$\frac{5}{100}$
  B one hundred two and—$\frac{50}{100}$
  C one thousand two and—$\frac{50}{100}$
  D one hundred two and—$\frac{5}{100}$

34 Which number sentence **does not** have an answer of 12?

  F $12 \times 1 = \square$
  G $12 + 1 = \square$
  H $12 - 0 = \square$
  J $12 \div 1 = \square$

35 Which of the following decimals, rounded to the nearest whole number, is 35?

  A 35.96
  B 34.94
  C 34.14
  D 35.67

This diagram shows how an artist plans to tile the bottom of a swimming pool. Study the diagram. Then do Numbers 36 through 38.

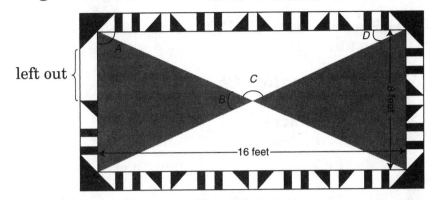

36 Look at the left side of the diagram. A section of the pool border has been left out on that side. Which of the figures below shows the section that is missing.

  F ▪   **G** ◣   **H** ◿   **J** ◤

37 Which of the angles labeled in the diagram is a right angle?

  A ∠A
  B ∠B
  C ∠C
  D ∠D

38 What is the area of the space within the border?

  F   24 square feet
  G   48 square feet
  H 128 square feet
  J   64 square feet

Pre-Test Evaluation Chart

Use the key to check your answers on the Skills Inventory Pre-Test. The Evaluation Chart shows where you can turn in the book to find help with the problems you missed.

Key

1	A
2	J
3	C
4	J
5	C
6	H
7	B
8	F
9	A
10	K
11	B
12	F
13	C
14	G
15	C
16	J
17	C
18	F
19	C
20	H
21	D
22	J
23	B
24	F
25	D
26	G
27	B
28	J
29	C
30	F
31	B
32	H
33	B
34	G
35	B
36	F
37	A
38	H

Evaluation Chart

Problem Numbers	Skill Areas	Practice Pages
2,7	Addition of Whole Numbers	9–20
3,13	Subtraction of Whole Numbers	21–31
1,4	Multiplication of Whole Numbers	32–41
5,6	Division of Whole Numbers	42–55
10, 12	Decimals	56–64
8, 9, 11, 30	Fractions, Ratios, Proportions	65–83
20, 21, 33, 34	Numeration Number Theory	1–8, 9, 21, 32, 42, 56
26, 27, 28, 29	Data Interpretation	84–96
14, 15, 18, 24	Pre-Algebra/ Algebra	97–109
16, 19, 32	Measurement	110–126
36–38	Geometry	127–140
17, 22, 25	Computation in Context	17–18, 28–29, 38–39, 52–53, 62, 77, 80
23, 31, 35	Estimation/ Rounding	5, 6, 16, 27, 37, 51, 76

Number Power Intermediate 1, Level M

Correlations Between Number Power Intermediate *and TABE*™ Mathematics Computation

Addition of Whole Numbers   **Pre-Test Score** ☐   **Post-Test Score** ☐

Subskill	TABE, Form 7 Item Numbers	TABE, Form 8 Item Numbers	Practice and Instruction Pages in This Text (*p* means practice page.)	Additional Practice and Instruction Resources
Adding 1–3 digits without regrouping		2	9–13, 19–20*p*	*Number Sense*, Bk. 1, pages 10–12, 14, 16, 17 *Breakthroughs in Math / Bk. 1*, pages 18–21 Number Power, Bk. 1, pages 6–8
Adding 4 digits without regrouping	3, 8		9–13, 19–20*p*	*Breakthroughs in Math / Bk. 1*, pages 24–27 *Number Power, Bk. 1*, pages 8–9
Adding 1–3 digits with regrouping	13	4	14–15, 19–20*p*	*Number Sense*, Bk. 1, pages 19–21 *Breakthroughs in Math / Bk. 1*, pages 32–34 *Number Power*, Bk. 1, pages 10–12
Adding 4 digits with regrouping	4	8	14–15, 19–20*p*	*Number Sense*, Bk. 1, pages 22–24 *Breakthroughs in Math / Bk. 1*, pages 33–37 *Number Power*, Bk. 1, pages 13–14
Adding three 1–digit numbers		1	13, 19–20*p*	*Breakthroughs in Math / Bk. 1*, pages 22–23 *Foundations: Mathematics*, pages 25–26

Corresponds to TABE*™ *Forms 7 and 8.
Tests of Adult Basic Education are published by CTB Macmillan/McGraw-Hill. Such company has neither endorsed nor authorized this test preparation book.

xiii

Subtraction of Whole Numbers　　Pre-Test Score ☐　　Post-Test Score ☐

Subskill	TABE, Form 7	TABE, Form 8	Practice and Instruction Pages in This Text	Additional Practice and Instruction Resources
Subtracting from 2 or 3 digits without regrouping	2	3	21–24, 30–31p	*Number Sense*, Bk. 1, pages 27–32 *Breakthroughs in Math/Bk. 1*, pages 44–48 *Number Power*, Bk. 1, pages 22–24
Subtracting from 4 digits without regrouping	10, 12		23–24, 30–31p	*Breakthroughs in Math/Bk. 1*, pages 48–50 *Number Power*, Bk. 1, pages 24–25
Subtracting from 2 or 3 digits with regrouping	5	13	25–26, 30–31p	*Number Sense*, Bk. 1, pages 33–42, 44, 45 *Breakthroughs in Math/Bk. 1*, pages 56–60, 62–64 *Number Power*, Bk. 1, pages 26–28, 32–33, 36
Subtracting from 4 digits with regrouping		10, 14	25–26, 30–31p	*Number Sense*, Bk. 1, pages 43–45 *Breakthroughs in Math/Bk. 1*, pages 57–64 *Number Power*, Bk. 1, pages 28–34, 36

Multiplication of Whole Numbers　　Pre-Test Score ☐　　Post-Test Score ☐

Subskill	TABE, Form 7	TABE, Form 8	Practice and Instruction Pages in This Text	Additional Practice and Instruction Resources
Multiplying by 1 digit without regrouping	1, 11, 18	6, 12	32–34, 40–41p	*Number Sense*, Bk. 2, pages 1–7, 9 *Breakthroughs in Math/Bk. 1*, pages 72–76, 81 *Number Power*, Bk. 1, pages 45–49, 52
Multiplying by 2 or more digits without regrouping		5	35, 40–41p	*Breakthroughs in Math/Bk. 1*, pages 77, 79, 81 *Number Power*, Bk. 1, pages 50–53
Multiplying by 1 digit with regrouping		23	36, 40–41p	*Number Sense*, Bk. 2, pages 10–12 *Breakthroughs in Math/Bk. 1*, pages 84–87, 91 *Number Power*, Bk. 1, pages 54–55
Multiplying by 2 or more digits with regrouping	7, 14	15	36, 40–41p	*Number Sense*, Bk. 2, pages 14–18 *Breakthroughs in Math/Bk. 1*, pages 88–91 *Number Power*, Bk. 1, pages 56–59

Division of Whole Numbers   Pre-Test Score ☐   Post-Test Score ☐

Subskill	TABE, Form 7	TABE, Form 8	Practice and Instruction Pages in This Text	Additional Practice and Instruction Resources
Basic division facts		9	42–43, 54–55p	*Number Sense*, Bk. 2, pages 23–25 *Breakthroughs in Math / Bk. 1,* pages 100–103
Dividing by 1 digit with no remainder	6, 9	7	44–46, 54–55p	*Number Sense*, Bk. 2, pages 26, 28, 30, 33 *Breakthroughs in Math / Bk. 1,* pages 104–107, 110–112 *Number Power*, Bk. 1, pages 72–75, 89
Dividing by 2 or more digits with no remainder	16	11	50, 54–55p	*Breakthroughs in Math / Bk. 1,* pages 120–121, 123 *Number Power*, Bk. 1, pages 80–82, 84, 86
Dividing by 1 digit with a remainder	17	16	47–49, 54–55p	*Number Sense*, Bk. 2, pages 27–29, 31–32 *Breakthroughs in Math / Bk. 1,* pages 108–112, 116–117 *Number Power*, Bk. 1, pages 76–79

Decimals   Pre-Test Score ☐   Post-Test Score ☐

Subskill	TABE, Form 7	TABE, Form 8	Practice and Instruction Pages in This Text	Additional Practice and Instruction Resources
Adding decimals	25		58, 63–64p	*Number Sense*, Bk. 3, pages 19–28 *Breakthroughs in Math / Bk. 2,* pages 42–43 *Pre-GED Satellite Program: Mathematics*, pages 53–56
Subtracting decimals		24, 25	59, 63–64p	*Number Sense*, Bk. 3, pages 33–41, 44 *Breakthroughs in Math / Bk. 2,* pages 44–45 *Pre-GED Satellite Program: Mathematics*, pages 53, 54, 56
Multiplying decimals	19, 22, 24	17, 21	60–61, 63–64p	*Number Sense*, Bk. 4, pages 1–17 *Breakthroughs in Math / Bk. 2,* pages 48–51 *Pre-GED Satellite Program: Mathematics*, pages 57–60

Fractions   **Pre-Test Score** ☐   **Post-Test Score** ☐

Subskill	TABE, Form 7	TABE, Form 8	Practice and Instruction Pages in This Text	Additional Practice and Instruction Resources
Adding fractions		18, 20	67, 71, 82–83*p*	*Number Sense,* Bk. 6, pages 1–4, 6–10 *Breakthroughs in Math / Bk. 2,* pages 66–69, 73, 74 *Number Power, Bk. 2,* pages 7–9, 11, 13–15
Subtracting fractions	15, 23		67, 72, 82–83*p*	*Number Sense,* Bk. 6, pages 27–29 *Breakthroughs in Math / Bk. 2,* pages 66–69, 76, 78 *Number Power, Bk. 2,* pages 7–9, 11, 21
Multiplying fractions	20, 21	19, 22	73–75, 82–83*p*	*Number Sense,* Bk. 7, pages 1–10 *Breakthroughs in Math / Bk. 2,* pages 66–69, 92, 94 *Number Power, Bk. 2,* pages 7–9, 11, 28, 31

Correlations Between Number Power Intermediate and TABE Applied Mathematics

Numeration   Pre-Test Score ☐   Post-Test Score ☐

Subskill	TABE, Form 7	TABE, Form 8	Practice and Instruction Pages in This Text	Additional Practice and Instruction Resources
Word names	23		2, 7–8p	*Number Sense,* Bk. 3, pages 9, 13–15 *Breakthroughs in Math / Bk. 1,* pages 9–11; *Bk. 2,* pages 34–38 *Pre-GED Satellite Program: Mathematics,* pages 14–16, 45–50
Recognizing numbers	18	4	2, 4, 7–8p	*Number Sense,* Bk. 1, pages 7–9; Bk. 3, pages 1–6, 12, 16–18; Bk. 5, pages 1–10 *Breakthroughs in Math / Bk. 1,* pages 8–11; *Bk. 2,* pages 34–39, 66–67, 72 *Pre-GED Satellite Program: Mathematics,* pages 11, 12, 14, 16, 45–50, 69–72
Ordering		24, 40	3, 7–8p, 57, 64p, 69–70, 82–83p	*Number Sense,* Bk. 1, pages 6–9; Bk. 3, pages 1–6, 16–18; Bk. 5, pages 14–17, 25–27 *Pre-GED Satellite Program: Mathematics,* pages 45–50, 81–83 *Foundations: Mathematics,* pages 4–7, 76, 123–124, 127, 129
Place value	6		1, 7–8p, 56, 64p	*Number Sense,* Bk. 1, pages 1–4, 7; Bk. 3, 1–16 *Breakthroughs in Math / Bk. 1,* pages 8, 9; *Bk. 2,* pages 34–36 *Foundations: Mathematics,* pages 3–7, 101–103
Comparison	19		3, 7–8p, 57, 64p, 69–70, 82–83p	*Number Sense,* Bk. 1, pages 7–9; Bk. 3, pages 1–6, 16–17; Bk. 5, pages 14–16, 25–27 *Number Power Review,* pages 2–4
Fractional part	16, 29	25, 48	65–67, 82–83p	*Number Sense,* Bk. 5, pages 1–13, 28–30, 55–59 *Breakthroughs in Math / Bk. 2,* pages 66–69, 72 *Number Power,* Bk. 2, pages 5–9

Correlation Tables

Number Theory   **Pre-Test Score** ☐   **Post-Test Score** ☐

Subskill	TABE, Form 7	TABE, Form 8	Practice and Instruction Pages in This Text	Additional Practice and Instruction Resources
Sequence	2, 9	1	97–102, 108–109p	*Critical Thinking with Math,* pages 3–5, 8–12 *Number Power Review,* pages 98–99
Properties	8	12	9, 20p, 21, 31p, 32, 41p, 42	*Number Sense,* Bk. 1, pages 10, 27, 30, 49; Bk. 2, pages 1–4, 21–24, 46 *Breakthroughs in Math / Bk. 1,* pages 18–19, 44–45, 72–73, 100–101 *Foundations: Mathematics,* pages 14, 17, 19, 30, 31, 39–40, 42, 44
Equivalent form	3, 33	28	4, 7–8p, 67	*Breakthroughs in Math / Bk. 1,* pages 14–15; *Bk. 2,* pages 68–70 *Number Power,* Bk. 2, pages 8–12, 49, 52–53 *Foundations: Mathematics,* pages 5–6, 75–81, 104–106, 129–135
Divisibility		13		*Number Sense,* Bk. 5, page 31–35
Factors		8	68	*Number Sense,* Bk. 5, page 31–41
Ratio, Proportion	43	20, 31	78–83	*Number Sense,* Bk. 8, page 1–60 *Foundations: Mathematics,* pages 151–158 *Pre-GED Satellite Program: Mathematics,* pages 149–160

Data Interpretation   **Pre-Test Score** ☐   **Post-Test Score** ☐

Subskill	TABE, Form 7	TABE, Form 8	Practice and Instruction Pages in This Text	Additional Practice and Instruction Resources
Graphs	14		88–94, 95–96p	*Number Power,* Bk. 5, pages 6–66; Bk. 8, pages 38–41, 44–45, 49–54, 60–61, 72–87, 96–97 *Real Numbers,* Bk. 3, pages 12–32, 46–48 *Pre-GED Satellite Program, Mathematics,* pages 191–199, 210–218
Tables, charts, and diagrams	20, 22, 28, 30, 31, 34, 36, 38	5, 6, 7, 10, 11, 15, 22, 23	84–89, 95–96p	*Number Power,* Bk. 5, pages 67–91; Bk. 8, pages 34–37, 60–61 *Real Numbers,* Bk. 3, pages 1–11, 35–39, 44–55 *Pre-GED Satellite Program, Mathematics,* pages 185–186, 200–209

Algebra   **Pre-Test Score** ☐   **Post-Test Score** ☐

Subskill	TABE, Form 7	TABE, Form 8	Practice and Instruction Pages in This Text	Additional Practice and Instruction Resources
Function/pattern	4, 35	2, 3, 42, 50	97–102, 108–109p	*Critical Thinking with Math,* pages 3–5, 8–12 *Number Power Review,* pages 98–99
Missing element	1, 7, 13	9	100–102, 108–109p	*Number Sense,* Bk. 1, pages 18, 30, 54, 55; Bk. 2, pages 8, 23 *Real Numbers,* Bk. 5, pages 5–10, 12–15 *Foundations: Mathematics,* pages 15, 16, 18
Strategy Application	12		103–107, 108–109p	*Critical Thinking with Math,* pages 16–22 *Number Power Review,* pages 172–179 *Real Numbers,* Bk. 5, pages 1–4, 11, 16, 20–21

Measurement   **Pre-Test Score** ☐   **Post-Test Score** ☐

Subskill	TABE, Form 7	TABE, Form 8	Practice and Instruction Pages in This Text	Additional Practice and Instruction Resources
Appropriate instrument		18	110, 125–126p	*Real Numbers,* Bk. 4, page 1
Money		47	4, 7–8p	*Number Sense,* Bk. 3, pages 29, 30, 56, 57 *Foundations, Mathematics,* pages 74–97 *Breakthroughs in Math/ Bk. 1,* pages 14–16
Time	5, 39, 47	43	123–124, 125–126p	*Real Numbers,* Bk. 4, pages 54–60, 64–67 *Number Power,* Bk. 9, pages 126–137, 140, 141
Temperature		21	111–114, 125–126p	*Real Numbers,* Bk. 4, pages 2–4, 7, 8, 46–52 *Number Power,* Bk. 9, pages 12, 13, 74–77
Mass, Weight		17, 19, 41	111–113, 121, 125–126p	*Real Numbers,* Bk. 4, pages 26–31 *Number Power,* Bk. 9, pages 52–65, 70–73
Perimeter	26, 45	36	119, 125–126p	*Real Numbers,* Bk. 6, pages 26–32, 42–45 *Breakthroughs in Math/ Bk. 1,* pages 154–155 *Number Power,* Bk. 4, pages 70–93
Area	25		120, 125–126p	*Real Numbers,* Bk. 6, pages 33–37, 42–45 *Breakthroughs in Math/ Bk. 1,* pages 156–157 *Number Power,* Bk. 4, pages 70–93

Geometry **Pre-Test Score** ☐ **Post-Test Score** ☐

Subskill	TABE, Form 7	TABE, Form 8	Practice and Instruction Pages in This Text	Additional Practice and Instruction Resources
Symmetry		32	135, 139–140p	*Real Numbers,* Bk. 6, page 63
Pattern, shape	50	34, 49	97–98, 108–109p	*Critical Thinking with Math,* pages 3–5, 8–12 *Number Power Review,* pages 98–99
Geometric elements	11		128, 130–133, 139–140p	*Real Numbers,* Bk. 6, pages 1, 2, 11–15 *Pre-GED Satellite Program, Mathematics,* pages 117–120
Congruency		35	136, 139–140p	*Real Numbers,* Bk. 6, pages 60–61
Plane figures	44		128–137, 139–140p	*Real Numbers,* Bk. 6, pages 1, 16–20 *Number Power,* Bk. 4, pages 82–83 *Pre-GED Satellite Program, Mathematics,* pages 117–120
Solid figures	49		134, 138, 139–140p	*Number Power,* Bk. 4, pages 120–121
Logical reasoning	37		127, 139–140p	
Angles		33	128–129, 139–140p	*Real Numbers,* Bk. 6, pages 3–8 *Number Power,* Bk. 4, pages 6–17 *Pre-GED Satellite Program, Mathematics,* pages 117–120
Similarity	48		137, 139–140p	*Real Numbers,* Bk. 6, pages 64–65
Parts of a circle		37	133, 139–140p	*Real Numbers,* Bk. 6, pages 46 *Number Power,* Bk. 4, pages 102

Computation in Context   **Pre-Test Score** ☐   **Post-Test Score** ☐

Subskill	TABE, Form 7	TABE, Form 8	Practice and Instruction Pages in This Text	Additional Practice and Instruction Resources
Whole Numbers	15, 40, 41	30	18–20, 28–31, 38–41, 44p, 47p, 48p, 52–55, 86, 88–89, 93p, 95–96p, 103–105	*Number Sense,* Bk. 1, pages 25–26, 46–60 *Number Power,* Bk. 6, pages 7–35, 55–69, 97–106 *Breakthroughs in Math/Bk. 1,* pages 28–31, 38–41, 51–55, 65–69, 82–83, 94–97, 113–115, 125–131, 134–139
Decimals	21, 24, 32	29, 38, 45	4, 60–61, 63–64p, 87	*Number Sense,* Bk. 3, pages 29–32, 45–54, 56–59 *Number Power,* Bk. 6, pages 36–47 *Breakthroughs in Math/Bk. 2,* pages 46–47, 58–63
Fractions		27	66p, 77–83, 91p, 92p, 94, 95–96p	*Number Sense,* Bk. 6, pages 47–55 *Number Power,* Bk. 6, pages 48–51 *Breakthroughs in Math/Bk. 2,* pages 80–81, 88–91, 107

Estimation   **Pre-Test Score** ☐   **Post-Test Score** ☐

Subskill	TABE, Form 7	TABE, Form 8	Practice and Instruction Pages in This Text	Additional Practice and Instruction Resources
Reasonableness of Answer	27, 42	39	5, 7–8p	*Real Numbers,* Bk. 1, pages 1, 2, 4, 16, 23, 30, 45, 50, 52, 59, 68 *Real Numbers,* Bk. 2, pages 1, 16, 23, 33, 37
Rounding	17, 46	14, 16, 44	6, 7–8p, 76	*Real Numbers,* Bk. 1, pages 1, 9–11, 17, 24–26, 39, 40, 43, 46, 53–56 *Real Numbers,* Bk. 2, pages 10, 11, 18, 21, 22, 25–27, 29, 30, 35, 36, 38 *Breakthroughs in Math/Bk. 1,* pages 12, 13
Estimation	10	26, 46	5, 7–8p, 16, 27, 37, 51	*Breakthroughs in Math/Bk. 1,* pages 92–95 *Number Power Review,* pages 8–12, 68, 69 *Real Numbers,* Bk. 1, pages 1, 3, 5–15, 17–22, 24–29, 31, 32, 34–38, 40–44, 46–49, 51, 53–58, 60, 62–69

The Number System

Finding Place Value

> The ten **digits** are 0, 1, 2, 3, 4, 5, 6, 7, 8, and 9. The value of a digit in a number depends on its **place** in that number.

Look at the numbers 17 and 71. They have the same digits, but they are different numbers. That is because the digits are in different places. The number 17 stands for **10 + 7** or **1 ten and 7 ones.** The number 71 stands for **70 + 1** or **7 tens and 1 one.**

```
ten thousands
      thousands
            hundreds
                  tens
                      ones
___  1 ,  4    0    3
```

The diagram above shows the first five places in whole numbers. The number 1,403 has digits in the first four places. It has 1 thousand, 4 hundreds, 0 tens, and 3 ones.

PRACTICE

Fill in the blanks.

1 472 has 4 _____, 7 _____, and 2 _____.

2 1,230 has _____.

3 31,405 has _____.

4 Look at the number 4,671. What place is the 6 in? _____

5 Look at the number 10,426. What place is the 1 in? _____

6 Look at the number 52,310. What digit is in the hundreds place? _____

7 Look at the number 76,802. What digit is in the thousands place? _____

8 In the number 8,312, the digit 8 is in the thousands place. Its value is eight thousand or 8,000. What is the value of the 3?

   _____

9 What is the value of the 5 in 25,036? _____

10 What is the value of the 6 in 45,609? _____

Naming Large Numbers

When you say or write a number in words, you use the place values of the digits.

784 =	700	+	80	+	4
= seven hundred			eighty-four		

The digits in the thousands and ten thousands places are grouped together.

41,032 =	41,000		+	30	+	2
= forty-one thousand,				thirty-two		

Notice that the name "forty-one thousand, thirty-two" does not contain "zero hundreds." If a number has a zero, you do not say the place name for that digit.

PRACTICE

Write each number in words.

1 3,482 _____

2 21,601 _____

3 4,003 _____

4 1,041 _____

5 23,010 _____

Write each number below in digits. Watch for number names that skip a place. You must put a zero in that position.

> two hundred six = 206
>
> Write "two hundred six" as "206." It is not "26" or "260."

6 nine hundred two _____

7 one thousand, two hundred sixty-three _____

8 fifty-two thousand _____

9 eighty thousand, one hundred forty-two _____

10 four thousand, three hundred six _____

11 thirteen thousand, six hundred _____

12 twenty-one thousand, five hundred eighty-five _____

Comparing Whole Numbers

If two whole numbers have different numbers of digits, the number with more digits is larger.
    5 , 0 1 3 is larger than   9 8 4

If two whole numbers have the same number of digits, start at the left to compare them.
    5 , 8 6 2 is larger than   2 , 9 7 4
because 5 is larger than 2.

PRACTICE

Circle the larger number in each problem.

1	2,842	546
2	193	20
3	21,403	899
4	89	98
5	876	413

Circle the largest number in each problem.

6	20,000	8,000	16,000
7	51	9	13
8	11,000	719	3,230
9	160	98	29
10	13	9	5

Arrange these numbers from smallest to largest.

11 97, 115, 146, 52

————, ————, ————, ————

12 103, 95, 17, 9

————, ————, ————, ————

The two numbers 4,917 and 4,958 have the same number of places. Starting at the left, both numbers have the digit 4, then both numbers have the digit 9. The next pair of digits is 1 and 5, so 4,958 is larger than 4,917.

13 178, 517, 203, 695, 961 ————, ————, ————, ————, ————

Arrange these digits to make the smallest number possible.

14 2, 1, 7 _____

15 8, 7, 6 _____

16 9, 2, 3 _____

Reading and Writing Dollars and Cents

Dollars and cents are written using a **decimal point.** (You will be learn more about decimals later in this book.)

When a dollar amount is written in decimal form, it looks like this: $5.06 (five dollars and six cents). Dollars are shown to the left of the decimal point. Cents are shown to the right of the decimal point. When you say or write an amount in words, you use the word **and** to represent the decimal point.

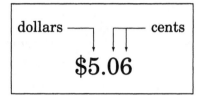

PRACTICE

Write each amount in words. *Hint:* **$0.05 is five cents, $0.50 is fifty cents, and $2.10 is two dollars and ten cents.**

1   $3.23 is _____ dollars and _____ cents.

2   $4.03 is _____ dollars and _____ cents.

3   $1.13 is _____ dollars and _____ cents.

4   $0.50 is _____ dollars and _____ cents.

5   $10.13 is _____ dollars and _____ cents.

6   $51.10 is _____ dollars and _____ cents.

7   $106.00 is _____ dollars.

> When you write an amount of money in decimal form, always use two digits to the right of the decimal point. For example, "two dollars and seven cents" is $2.07. Never write "$2.7" because it is not clear whether that means $2.70 or $2.07.

Write these amounts in decimal form.

8   eight dollars and
    fifty-nine cents   _____

9   two dollars and
    nine cents   _____

10 ten dollars and
    fifty cents   _____

11 eighty-two dollars
    and sixteen cents   _____

12 twenty-five cents   _____

13 97¢   _____

14 8¢   _____

Estimating

Sometimes you need to know *about how much money* something costs or *about how much time* something takes. This kind of amount is called an **estimate.** An estimate is a number that is close to an actual or exact amount.

All the words in the box signal that a math problem calls for estimation.

> *almost*
> *approximately*
> *about*

PRACTICE

1 **Circle the letters for the problems that call for estimation.**

A There are about fifty ham sandwiches and a little over ten cheese sandwiches in a freezer. About how many sandwiches are there in all?

B Lonnie began work at 8:15. He left at 4:52. How long was he at work?

C It takes 3 cups of flour to make a loaf of bread. You have 12 cups of flour. How many loaves of bread can you make?

D There are 52 boxes of peaches. Each box holds 26 cans. Approximately how many cans of peaches are there?

Often you can use common sense to come up with an estimate. For instance, you know from experience that a hamburger and an order of fries should cost about five dollars, not fifty dollars. This type of common sense is very useful when you are doing math problems. *Get into the habit of using it to check whether your answers make sense.*

Circle the letter for the answer that makes the most sense.

2 You buy 2 pounds of hamburger. About how much should it cost?

F $4.00
G $12.00
H $21.00
J $120.00

3 You drive a car 300 miles. About how long should it take?

A 1 hour
B 5 hours
C 20 hours
D 2 days

4 You buy a sandwich at a convenience store. You pay with a 10-dollar bill. About how much change should you get?

F 5 cents
G 10 cents
H 5 dollars
J 10 dollars

5 You buy a shirt, a pair of pants, and a pair of shoes. About how much should it cost?

A $150.00
B $15.00
C $500.00
D $1,000.00

Rounding

Another way to estimate the answer to a math problem is use rounded numbers. A rounded number is close to the exact one, but it is easier to work with.

To round a number, think of it on a hilly number line like this. Each low spot ends in a zero. Numbers ending in 0, 1, 2, 3, and 4 roll back to the nearest low spot. Numbers ending in 5, 6, 7, 8, and 9 roll ahead to the nearest low spot.

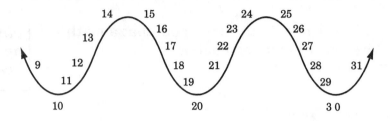

Examples:

11 rounds to 10.   20 rounds to 20.
18 rounds to 20.   5 rounds to 10.

PRACTICE

Circle the number that completes each of the statements below.

1	82 rounds to	90	80
2	117 rounds to	110	120
3	99 rounds to	90	100
4	$1.03 rounds to	$0.90	$1.00
5	$5.14 rounds to	$5.10	$5.20
6	1,903 rounds to	1,900	1,910

Round each number to the nearest ten.

7   78 rounds to _____.

8   102 rounds to _____.

9   21 rounds to _____.

10 $1.04 rounds to _____.

11   99 cents rounds to _____.

12   18 rounds to _____.

13   86 rounds to _____.

You can round a number to different place values.

◆ 191 rounded to the tens place is 190.
◆ 191 rounded to the hundreds place is 200.

To round a number to a place value, look at the digit just to the right of that place value. If that digit is less than 5, round down. If that digit is 5 or more, round up.

14   Round 115 to
the hundreds place.   _____

15   Round 2,123 to
the tens place.   _____

16   Round 51,620 to
the thousands place.   _____

17   Round 867 to
the tens place.   _____

18   Round 915 to
the thousands place.   _____

19   Round $1.98 to
the nearest dollar.   _____

Number System Skills Practice

Circle the letter for the correct answer to each problem.

1 What is the value of the digit 6 in the number 26,503?

   A   60
   B   600
   C   6,000
   D 60,000

2 Which of the following numbers is *greater* than 3,645?

   F 2,759
   G 3,367
   H 3,731
   J 3,639

3 Which of these numbers is 819 rounded to the nearest ten?

   A 900
   B 800
   C 810
   D 820

4 Which of these has the same value as 6 thousands, 5 hundreds, 2 ones?

   F   652
   G 6,502
   H 6,520
   J 6,052

5 You hear that the new city manager will earn one hundred twenty thousand dollars a year. Which of these numbers shows one hundred twenty thousand dollars?

   A $120,000
   B $12,000
   C $102,000
   D $1,020,000

6 Claude spent $58.15 on shoes and $137.80 on clothing. He wants to estimate the total amount he spent by rounding each purchase to the nearest dollar. Which pair of numbers should he use?

   F $58.00 and $140.00
   G $58.00 and $138.00
   H $60.00 and $140.00
   J $58.00 and $137.00

7 Which group of prices is in order from least to greatest?

   A $0.31, $0.70, $1.00, $1.50
   B $1.00, $1.50, $0.70, $0.31
   C $0.70, $0.31, $1.00, $1.50
   D $1.00, $1.50, $0.31, $0.70

Mr. and Mrs. Suarez just opened an ice cream shop. This list shows how many people visited the shop each of their first four days.

Friday	182
Saturday	167
Sunday	109
Monday	98

Use the list to answer questions 8 through 13.

8   On which day did the shop have the most customers?

   F Friday
   G Saturday
   H Sunday
   J Monday

9   Mrs. Suarez decides to estimate the total number of customers by rounding each number to the nearest ten. Which set of numbers should she use?

   A 100, 200, 100, 100
   B 180, 160, 100, 90
   C 180, 170, 100, 100
   D 180, 170, 110, 100

10   Mr. Suarez needs to give a customer 40 cents in change. Which of these groups of coins is worth 40 cents?

   F two quarters
   G three dimes and a nickel
   H three nickels and a quarter
   J five nickels

11   A customer purchases a cone for $1.03. What is that amount in words?

   A one dollar and thirty cents
   B thirteen dollars
   C thirteen cents
   D one dollar and three cents

12   For which of the following tasks could Mr. and Mrs. Suarez use an estimate?

   F finding out the price of three cones
   G finding out how much change a customer should receive
   H finding out how much they owe the ice cream factory
   J finding out how many customers to expect next week

13   Mr. and Mrs. Suarez are hiring a clerk to help out at the shop. At first, the clerk will be paid the minimum wage. Which of these is the most reasonable estimate of how much the clerk will earn in one 8-hour day?

   A $5.00
   B $50.00
   C $150.00
   D $95.00

Addition

Basic Concepts

The math term for putting things together is **adding.**

- These words signal an addition problem:
  five *plus* three   six *added to* three
  the *sum* of seven and two   how many *altogether*
  the *total* of one and three   how many *in all*

- An addition problem can be written from top to bottom in a **column** or from left to right in a **row.** The answer is called the **sum** or **total.**

  > column
  > row → 3 + 2 = 5   3
  >   + 2
  >   sum → 5

- You can change the order of the numbers when you add. That does not change the sum.

  > **2 + 5 has the same value as 5 + 2.**

- If you add zero to a number, you do not change the value.

  > **2 + 0 = 2**

- You can only add like things together.

  > You cannot add 4 hours and 6 airplanes.

- Addition and subtraction are the reverse of each other. To "undo" the result of adding 3, you can subtract 3.

  > **2 + 3 = 5 so 5 – 3 = 2**

PRACTICE

1 Write this addition problem two different ways: three plus six equals 9.

2 Which problem has the same value as 265 + 57?

   A 265 – 57
   B 57 – 265
   C 57 + 265
   D None of them

3 423 + 0 = _____

4 *True* or *false:* 2 flies + 3 ants = 5 flies

Basic Addition Facts

To solve problems using addition, you must know the basic facts quickly and accurately. If you make more than a few mistakes on these problems, erase your answers and try the problems again.

> (*Remember:* In a sum, the order of the numbers is not important.)

PRACTICE

1 2 + 3 = _____   3 + 3 = _____   7 + 7 = _____   0 + 3 = _____

2 8 + 8 = _____   6 + 4 = _____   2 + 2 = _____   9 + 6 = _____

3 1 + 1 = _____   6 + 5 = _____   6 + 3 = _____   1 + 3 = _____

4 5 + 2 = _____   4 + 7 = _____   0 + 0 = _____   4 + 1 = _____

5
$$\begin{array}{r} 4 \\ + 3 \\ \hline \end{array} \qquad \begin{array}{r} 7 \\ + 8 \\ \hline \end{array} \qquad \begin{array}{r} 4 \\ + 4 \\ \hline \end{array} \qquad \begin{array}{r} 8 \\ + 9 \\ \hline \end{array} \qquad \begin{array}{r} 9 \\ + 9 \\ \hline \end{array} \qquad \begin{array}{r} 4 \\ + 0 \\ \hline \end{array}$$

6
$$\begin{array}{r} 7 \\ + 0 \\ \hline \end{array} \qquad \begin{array}{r} 7 \\ + 5 \\ \hline \end{array} \qquad \begin{array}{r} 3 \\ + 5 \\ \hline \end{array} \qquad \begin{array}{r} 2 \\ + 4 \\ \hline \end{array} \qquad \begin{array}{r} 1 \\ + 2 \\ \hline \end{array} \qquad \begin{array}{r} 3 \\ + 0 \\ \hline \end{array}$$

7
$$\begin{array}{r} 5 \\ + 1 \\ \hline \end{array} \qquad \begin{array}{r} 6 \\ + 2 \\ \hline \end{array} \qquad \begin{array}{r} 1 \\ + 7 \\ \hline \end{array} \qquad \begin{array}{r} 8 \\ + 4 \\ \hline \end{array} \qquad \begin{array}{r} 6 \\ + 6 \\ \hline \end{array} \qquad \begin{array}{r} 3 \\ + 7 \\ \hline \end{array}$$

8
$$\begin{array}{r} 8 \\ + 2 \\ \hline \end{array} \qquad \begin{array}{r} 3 \\ + 8 \\ \hline \end{array} \qquad \begin{array}{r} 0 \\ + 2 \\ \hline \end{array} \qquad \begin{array}{r} 1 \\ + 8 \\ \hline \end{array} \qquad \begin{array}{r} 5 \\ + 4 \\ \hline \end{array} \qquad \begin{array}{r} 0 \\ + 5 \\ \hline \end{array}$$

9
$$\begin{array}{r} 9 \\ + 4 \\ \hline \end{array}$$
5 + 5 = _____
$$\begin{array}{r} 8 \\ + 7 \\ \hline \end{array}$$
6 + 1 = _____
$$\begin{array}{r} 8 \\ + 0 \\ \hline \end{array}$$

10
$$\begin{array}{r} 8 \\ + 5 \\ \hline \end{array}$$
9 + 3 = _____
$$\begin{array}{r} 7 \\ + 9 \\ \hline \end{array}$$
0 + 6 = _____
$$\begin{array}{r} 5 \\ + 8 \\ \hline \end{array}$$

Simple Addition

To add large numbers, such as 114 and 203, start by writing them in column form. Set up the problem like the example below. Notice that the digits line up in columns.

Start at the digits in the right column. Add those digits, and write the sum. Then move one column to the left and add the digits in that column. Keep this up until you have added all the columns.

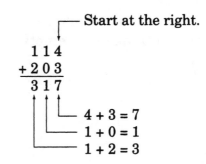

One way to check an addition problem is to cover the sum and add the numbers in each column again, from the bottom to the top.

PRACTICE

Find each sum. Then check your answer by adding the columns from the bottom to the top.

1	21	130	11	602
	+ 73	+ 251	+ 82	+ 215

> If the numbers in a problem have decimal points or labels, repeat them in your answer. In an addition problem, the decimal points line up vertically.

2	50	1,302	811	80
	+ 29	+ 5,251	+ 132	+ 15

3	105	630	660	5,210	57	62,998
	+ 203	+ 309	+ 12	+ 3,250	+ 22	+ 15,000

4	15 miles	711 pounds	500	47	1,200
	+ 44 miles	+ 105 pounds	+ 100	+ 41	+ 2,698

5	60	87 ft	31 cents	206	51,000
	+ 20	+ 11 ft	+ 52 cents	+ 391	+ 20,000

6	$5.00	555	$22.00	$12.00	92
	+ 2.03	+ 221	+ 31.00	+ 81.00	+ 23

7	$3.10	77 in.	1,104	$5.62	$3.30
	+ 1.14	+ 21 in.	+ 6,002	+ 1.20	+ 4.16

Adding Small and Large Numbers

To add numbers like 523 and 34, line up the digits in the ones column.

Then start at the right and add the columns. You can think of the empty space as a zero.

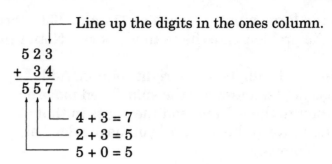

Line up the digits in the ones column.

```
  5 2 3
+   3 4
  5 5 7
```

4 + 3 = 7
2 + 3 = 5
5 + 0 = 5

PRACTICE

Find each sum. Then check your answer with common-sense estimation or bottom-to-top addition.

1	52	206	$83.00	1,010	12 feet
	+ 7	+ 82	+ 506.00	+ 500	+ 6 feet

2	418	8,000	45 in.	30,000	516 gal
	+ 80	+ 63	+ 541 in.	+ 1,300	+ 80 gal

Rewrite each problem in column form. Be sure to line up the digits in the ones place. Find each sum and check your answer.

3 51 + 7 = _____

6 670 + 4,015 = _____

9 113 + 6,375 = _____

4 5 + 71 = _____

7 517 + 40 = _____

10 What is 89 minutes plus 110 minutes?

5 $10.00 + $9.00 = _____

8 $6.00 + $810.00 = _____

Adding Three or More Numbers

You always add two digits at a time—even when there are three or more numbers to add. To add several numbers together, start by finding the sum of the first two digits. Then add the next digit to that sum. Keep adding the "next" digit until you complete the column. Then go on to the next column of digits.

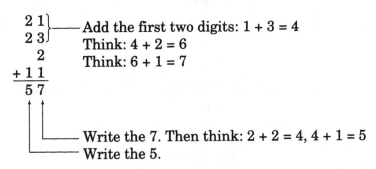

PRACTICE

Solve each problem. Check whether or not your answers seem correct.

1
$$\begin{array}{r} 11 \\ 20 \\ + 32 \\ \hline \end{array}$$
$$\begin{array}{r} 13 \\ 5 \\ + 1 \\ \hline \end{array}$$
$$\begin{array}{r} 2 \\ 3 \\ + 5 \\ \hline \end{array}$$
$$\begin{array}{r} 4¢ \\ 12¢ \\ + 2¢ \\ \hline \end{array}$$
$$\begin{array}{r} 16 \\ 12 \\ + 11 \\ \hline \end{array}$$
$$\begin{array}{r} 224 \\ 52 \\ + 11 \\ \hline \end{array}$$

2
$$\begin{array}{r} 131 \\ 14 \\ + 22 \\ \hline \end{array}$$
$$\begin{array}{r} 25 \text{ yards} \\ 12 \text{ yards} \\ + 2 \text{ yards} \\ \hline \end{array}$$
$$\begin{array}{r} 42 \\ 53 \\ + 41 \\ \hline \end{array}$$
$$\begin{array}{r} 7 \\ 3 \\ + 1 \\ \hline \end{array}$$
$$\begin{array}{r} 4 \\ 2 \\ + 6 \\ \hline \end{array}$$
$$\begin{array}{r} 8 \text{ in.} \\ 2 \text{ in.} \\ + 5 \text{ in.} \\ \hline \end{array}$$

3   12 + 4 + 41 = _____   15 + 20 + 20 = _____

4   60 + 30 + 80 = _____   71 + 26 + 2 = _____

5   32 + 53 + 24 = _____   19 + 100 + 60 = _____

6   61 + 103 + 2 = _____   16 + 12 + 81 = _____

Adding a Column When the Sum Is More than 9

In the problem below, the sum of the digits in the ones column is the two-digit number 14. For this problem, write the 4 in the ones place. Then "carry" the 1 to the tens column, and add all the digits in the tens column.

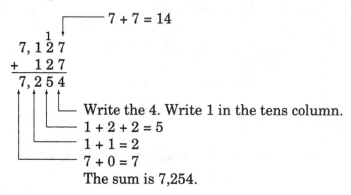

7 + 7 = 14

Write the 4. Write 1 in the tens column.
1 + 2 + 2 = 5
1 + 1 = 2
7 + 0 = 7
The sum is 7,254.

In this problem, the "carry" digit goes to the hundreds column.

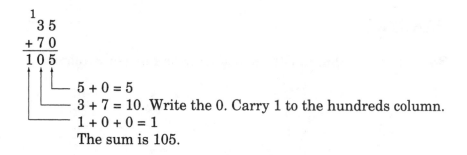

5 + 0 = 5
3 + 7 = 10. Write the 0. Carry 1 to the hundreds column.
1 + 0 + 0 = 1
The sum is 105.

PRACTICE

Find each sum. Then check your work.

1
59	88	55
+ 7	+ 30	+ 15

> If an addition problem has decimal points, be sure they line up vertically. After you add the columns, write a decimal point in your answer. It should line up with the other decimal points.

2
125	1,050	69 cm
+ 65	+ 75	+ 19 cm

3
68 gal	10,602	99 cans	59 in.	90
+ 203 gal	+ 900	+ 8 cans	+ 213 in.	+ 30

4
32 miles	5,609	134	649 in.	761
+ 193 miles	+ 919	+ 82	+ 270 in.	+ 57

5
290	$15.50	$29.00	150	657	319
+ 230	+ 25.00	+ 52.00	+ 90	+ 75	+ 4

In the problem at the right, the sum of the digits in both the ones column and the tens column is more than 9. This problem involves two "carry" digits.

```
  1 1
  1 5 9
+ 2 7 3
-------
  4 3 2
```

9 + 3 = 12. Write the 2 and carry the 1.
1 + 5 + 7 = 13. Write the 3 and carry the 1.
1 + 1 + 2 = 4

Rewrite each problem in column form. Be sure to line up the digits in the ones column. Find each sum and check your answer.

6   $17.00 + $13.00 = _____

7   146 + 82 = _____

8   $1.23 + $0.67 = _____

9   612 + 39 = _____

10   700 + 500 = _____

11   $42.00 + $83.00 = _____

12   $2.25 + $0.15 = _____

13   2,405 + 705 = _____

14   55 + 475 = _____

15   $6.50 + $1.75 = _____

16   163 + 459 = _____

17   365 + 87 = _____

18   2,670 + 2,430 = _____

19   $27.15 + $40.97 = _____

Using Estimation To Check Addition

You can check addition by using common sense and by adding columns from bottom to top. another method uses the first digit in each number.

Add the digits in the left column. Write a zero in each of the other columns.

┌─ Add the digits in the left column.
↓

16,592
+85,186
90,000

└─ Write zeros in all the other columns. The estimate is 90,000.

This method is called **front-end estimation.** With this method, the exact answer will always be **more than** your estimate.

PRACTICE

Estimate each sum using the first digit of each number. *Remember:* **With front-end estimation, your estimate should have at least as many places as the largest number in the problem.**

1
7,312	5,527	91 yards	$66.89
+ 6,903	+ 6,301	+ 73 yards	+ 12.43

2
549	20,212	1,893	257
+ 315	+ 13,500	+ 4,615	+ 13

Circle the letter for the *best* estimate for each problem.

3
115
+ 617

A   600
B   exactly 700
C   less than 700
D   more than 700

4   6,785 + 4,609 = _____

F   6,000
G   4,000
H   more than 10,000
J   less than 10,000

5   $3.97 + $5.12 = _____

A   less than $3.00
B   more than $8.00
C   less than $5.00
D   less than $8.00

6
1,052
+ 359

F   1,000
G   2,000
H   4,000
J   900

7
6,840
+ 1,193

A   more than 7,000
B   more than 70,000
C   less than 7,000
D   exactly 70,000

8   $5.63 + $8.19 = _____

F   less than $13.00
G   more than $1.30
H   more than $13.00
J   less than $1.30

Solving Word Problems

Here are five steps to solving a word problem.

1. Identify the particular question in the problem.
2. Get all the information you need.
3. Set up the problem.
4. Solve it.
5. Check your work.

You have already had practice with the first step. Remember that the signal words *plus, sum, total, added to, altogether,* and *in all* usually call for addition.

PRACTICE

1 **Circle the letter for the problems that call for addition. Do not try to solve the problems.**

A Quinton just bought a bed that is 7 feet long. His bedroom is 10 feet long. If he puts the bed into the corner, how much space will be left between the end of his bed and the wall?

B At rush hour, the Busy Bee cafe uses 4 waitresses, 2 cooks, 2 busboys, 1 manager, and 1 dishwasher. How many people in all work at the Busy Bee during rush hour?

C It costs 50 cents to play a game of "Paratroopers Attack!" How many games can you play for $5.00?

D Two companies share the same office building. One company has 14 employees. The other has 52 employees. Each employee gets one space in the parking lot. The parking lot must have at least how many spaces?

E There will be 2 speakers at the banquet. If each speaker talks for 25 minutes, at least how long will the banquet last?

F The speed limit is 40 miles an hour. Alex is driving 25 miles an hour. How far under the speed limit is he driving?

G People all over Renee's neighborhood have started putting ceramic geese in their yards. Renee counted 12 geese on her block. There are another 9 geese on the next block. How many geese is that altogether?

H Monte bought a cheese cake for $8.15 and a coffee cake for $6.98. How much did she spend in all?

I Tori won $25.00 playing bingo, but she spent $7.50 on bingo cards. How much money did she have after she subtracted the cost of the bingo cards?

Use the information below to set up problems 2–5. Then solve them. Check your work using the first digit of each number and front-end estimation.

> The dining room at the Busy Bee Cafe seats 250 people. There are seats for another 23 people at the counter. Today's specials are the hot beef sandwich for $4.45 and meatloaf dinner for $5.50.

2 If every seat is filled, how many people can sit at the Busy Bee?

4 Tax on the meatloaf dinner is $0.25. How much would it cost altogether to pay for a meatloaf dinner?

3 How much would it cost to buy two hot beef sandwiches? (Ignore the tax.)

5 The salad bar costs $2.50 extra. How much would it cost to buy the meatloaf dinner with the salad bar? (Ignore the tax.)

In some math problems, you have to find some missing information before you can solve it. On each blank lines, tell what information is needed to solve the problem.

6 Mr. Nguyen bought a shirt and spent $15.95 on gas. How much did he spend in all?

You also need to know _____

_____

7 Kali's cat gained 2 pounds this year. How much does it weigh now?

You also need to know _____

_____

8 When Wendy bought her car, she paid an extra $850.00 for air conditioning. How much did the car cost?

You also need to know _____

_____

9 Frederick's insurance payment just went up $20. How much does he pay now?

You also need to know _____

_____

Addition Skills Practice

Circle the letter for the correct answer to each problem. To start each problem, try crossing out all unreasonable answers for that problem.

1
$$\begin{array}{r} 2{,}378 \\ + 1{,}321 \end{array}$$

A 3,679
B 3,099
C 3,659
D 3,699
E None of these

2
$$\begin{array}{r} 1{,}876 \\ + \quad 8 \end{array}$$

F 1,874
G 1,884
H 1,076
J 1,878
K None of these

3 $516 + 29 =$ _____

A 545
B 535
C 806
D 525
E None of these

4 $16 + 402 + 31 =$ _____

F 548
G 718
H 448
J 449
K None of these

5
$$\begin{array}{r} 218 \\ + 406 \end{array}$$

A 224
B 724
C 614
D 714
E None of these

Use this information to do Numbers 6 through 8.

Marshall is 62 miles from Lincoln.
Lincoln is 43 miles from Rush.
Rush is 39 miles from Calloway.

6 You must drive from Lincoln to Rush and then back again. How far will you drive?

F 62 miles
G 43 miles
H 45 miles
J 86 miles

7 Jordan is driving from Marshall to Lincoln to see his parents, then on to Rush to visit a friend. How long will his trip be?

A 62 miles
B 43 miles
C 105 miles
D 15 miles

8 This month, Sarah took two trips. One was to Marshall and back. The other was to Rush and back. If she lives in Lincoln, how many miles did these two trips put on her car?

F 105 miles
G 210 miles
H 148 miles
J 167 miles

9

    462
+ 163

A 525
B 625
C 615
D 626
E None of these

10

$15.00 + 3.51 =$ _____

F 50.10
G   5.10
H 45.51
J 18.51
K None of these

11

  $2.90
+ 1.85

A $4.75
B $3.75
C $4.70
D $4.85
E None of these

12

$5 + 5 + 9 =$ _____

F 9
G 19
H 15
J 18
K None of these

13

  80
+ 25

A 95
B 100
C 115
D 125
E None of these

14 $20¢ + \$1.45 + \$5.00 =$ _____

F $5.65
G $8.45
H $6.65
J $6.47
K None of these

15 Which of these, if any, has the same value as $10 + 9$?

A $1 + 0 + 9$
B $6 + 4 + 9$
C $10 + 3 + 3$
D None of these

16 Which of these *does not* have the same value as $52 + 29$?

F $29 + 52$
G   52
     + 29

H $52 + 29 + 0$
J $52 + 29 + 1$

17 Which of the following amounts is different from the others?

A 25 cents + 75 cents
B ten dimes
C 50 cents + 50 cents
D ten nickels and one quarter

18 Which of the following phrases represents "12 + 7" ?

F twelve groups of seven
G one-seventh of twelve
H twelve minus seven
J the total of twelve and seven

Subtraction

Basic Concepts

The math term for taking something away, or finding the difference between two amounts, is **subtracting.**

- These words and phrases signal a subtraction problem.

minus	*difference*
take away	*how much change*
left over	*how much more or less*

- A subtraction problem can be written in a row or in a column. The symbol for subtraction is "−." The answer in a subtraction problem is called the **difference.**

- If you subtract zero from a number, you do not change the value.

$$2 - 0 = 2$$

- If you subtract a number from itself, the difference is zero.

$$7 - 7 = 0$$

- You cannot change the order of the numbers in a subtraction problem.

$4 - 2$ and $2 - 4$ have different values.

- Addition and subtraction are the reverse of each other. To "undo" the result of subtracting 4, you can add 4.

$$9 - 4 = 5$$
$$\text{so } 5 + 4 = 9$$

- You can only subtract *like* things. For example, you cannot subtract 7 apples from 20 oranges.

PRACTICE
Write T for *true* or F for *false.*

1 567 − 98 has the same value as 98 − 567. _____

2 8,932 − 0 = 8,932 _____

3 367 − 367 = 0 _____

4 9 − 2 − 1 has the same value as 7 − 1. _____

5 The same number can go in both boxes:
If 8 − ☐ = 5, then
5 + ☐ = 8. _____

Basic Subtraction Facts

To solve problems using subtraction, you must know the basic facts quickly and accurately. If you make more than a few mistakes on these problems, erase your answers and try the problems again.

PRACTICE

1   8 – 6 = _____   7 – 3 = _____   10 – 5 = _____

2   6 – 4 = _____   10 – 2 = _____   9 – 6 = _____

3   10 – 1 = _____   6 – 5 = _____   6 – 3 = _____

4   5 – 2 = _____   7 – 4 = _____   3 – 2 = _____

> *Hint:* If you know the basic addition facts, you can figure out the subtraction facts.
> *Example:* From
> $$2 + 3 = 5$$
> you can conclude that
> $5 - 3 = 2$ and $5 - 2 = 3$

5

4	8	4	9	9	4
−3	−7	−4	−8	−3	−0

6

10	7	5	4	2	3
−4	−5	−3	−2	−1	−0

7

5	6	7	8	7	10
−1	−2	−1	−4	−6	−3

8

8	8	10	8	5	5
−2	−3	−2	−1	−4	−0

9

9	5 – 5 = _____	10	6 – 1 = _____	8
−4		−7		−0

10

8	9 – 3 = _____	9	10 – 6 = _____	9
−5		−7		−5

Subtracting Larger Numbers

To subtract numbers such as 352 − 322 = _?_ , write the problem in column form. The number you are subtracting *from* must be on top.

Starting at the right, subtract the digits in each column. If the difference in a column is zero, be sure to write a zero below the column. However, you *do not* write a zero at the left of a whole number.

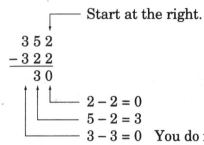

Start at the right.

$$\begin{array}{r} 3\,5\,2 \\ -\,3\,2\,2 \\ \hline 3\,0 \end{array}$$

2 − 2 = 0
5 − 2 = 3
3 − 3 = 0   You do not write a zero at the left of a whole number.

You can use addition to check a subtraction problem. Add the difference (the bottom number) to the number being subtracted (the middle number). The result should be the top number. This is an example that addition and subtraction "undo" each other.

Problem:

$$\begin{array}{r} 3\,5\,2 \\ -\,3\,2\,2 \\ \hline 3\,0 \end{array}$$ OK

Check:

$$\begin{array}{r} 3\,2\,2 \\ +\ \ 3\,0 \\ \hline 3\,5\,2 \end{array}$$

PRACTICE

Rewrite each problem in column form. Solve the problem, and then check your work.

1   870 − 620 = _____

2   462 − 231 = _____

3   7,910 − 6,500 = _____

4   231 miles − 220 miles = _____

5   550 − 240 = _____

6   1,302 − 1,100 = _____

7   5,400 − 3,100 = _____

> If the numbers in a problem have labels or decimal points, repeat them in your answer. The decimal points in a subtraction problem line up vertically.

8   1,050 − 1,020 = _____

9   $3.45 − $1.32 = _____

10   1,940 − 1,820 = _____

11   $25.00 − $15.00 = _____

12   86 in. − 32 in. = _____

13   $17.89 − $15.62 = _____

In the problem 289 – 23 = __?__ , the numbers have different numbers of digits. To start, line up the numbers in the ones column. If there is an empty space in a column, think of it as a zero.

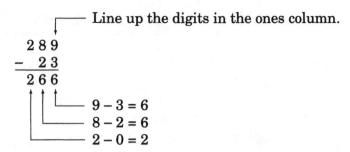

Line up the digits in the ones column.

```
  2 8 9
–   2 3
  2 6 6
```
9 – 3 = 6
8 – 2 = 6
2 – 0 = 2

Rewrite each problem in column form. Be sure to line up the digits in the ones place. Then find the diference and check your work.

14   692 – 81 = _____

15   781 – 60 = _____

16   910 – 10 = _____

17   1,500 – 300 = _____

18   5,890 – 30 = _____

19   842 – 30 = _____

20   7.47 – 0.31 = _____

21 $8.80 – $0.10 = _____

22 $291 – $80 = _____

23   6,714 – 700 = _____

24   25.32 – 2.10 = _____

25   46 – 15 = _____

26   32 – 31 = _____

27   31,412 – 300 = _____

28   62,805 – 302 = _____

29   52.98 – 1.60 = _____

30   67.95 – 5.15 = _____

31   117.00 – 5.00 = _____

Borrowing

In this subtraction problem, 4 is larger than 3 in the ones column. The top number is 20 + 3, so think of it as 10 + 13. In this way, you are "borrowing" from the tens column and giving that value to the ones column.

In the problem at the right, you borrow from the hundreds column. When you borrow, **neatly** cross out the number you are borrowing from, and write a number above it that is **one less.** Then cross out the digit that is too small, and write a number above it that is **ten more.**

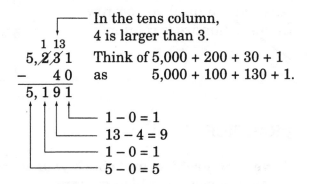

PRACTICE

Solve each problem. Then check your work.

1	82 − 8	**6**	1,500 meters − 7 meters		
2	142 − 15	**7**	1,940 feet − 903 feet		
3	618 − 26	**8**	$12.50 − 1.25		
4	$1.07 − 0.35	**9**	$15.50 −14.25	**11**	6,781 miles − 290 miles
5	917 − 891	**10**	890 −27	**12**	3,652 − 591

> If a subtraction problem has decimal points, be sure they line up vertically. After you subtract the columns, write a decimal point in your answer. It should line up with the other decimal points.

You need to borrow two times for this problem. Start on the right and do one column at a time. *Remember:* once you have crossed out a number, you must use the new number written above it.

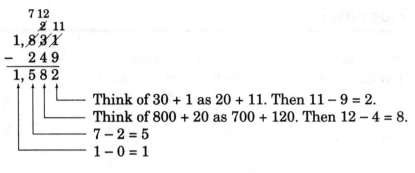

Think of 30 + 1 as 20 + 11. Then 11 − 9 = 2.
Think of 800 + 20 as 700 + 120. Then 12 − 4 = 8.
7 − 2 = 5
1 − 0 = 1

In the problem at the right you cannot borrow from the tens column until you borrow from the hundreds column.

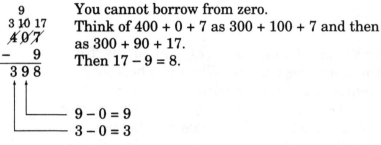

You cannot borrow from zero. Think of 400 + 0 + 7 as 300 + 100 + 7 and then as 300 + 90 + 17. Then 17 − 9 = 8.

9 − 0 = 9
3 − 0 = 3

PRACTICE

Solve each problem and check your work. Be sure to work neatly so it is clear which number goes with which column.

13   712
     − 29

14   911
     − 85

15   500
     − 76

16    $1.11
     − 0.32

17   1,000
     −   8

18   500 cm
     − 5 cm

19   3,000
     − 90

20   10,000
     −   51

21   918 − 99 = _____

22   4,000 − 25 = _____

23   3,100 − 56 = _____

Here is a shortcut to borrow from a row of zeros. Write a digit one less than the first nonzero digit. Then write 10 over the digit you need and write 9 over each digit in between.

```
   3 9 9 9 10
   4̶0̶,̶0̶0̶0̶
   −    5 2 8
   3 9, 4 7 2
```

24   314 − 75 = _____

25   6.00 − 0.25 = _____

26   12.50 − 5.75 = _____

Borrowing

Using Estimation To Check Subtraction

Here are two different ways to estimate a difference. One method, which uses the first digit in each number, is called **front-end estimation.**

Front-End Estimation
Subtract the digits in the left column.

$$\begin{array}{r} 6,9\,4\,2 \\ -\ 1,2\,4\,6 \\ \hline 5,0\,0\,0 \end{array}$$

Write zeros for all the other columns.

Another method uses rounding. In the problem at the right, the first step is to round each number to the thousands place. The second step is to subtract using the rounded numbers.

$$\begin{array}{r} 6,9\,4\,2 \\ -1,2\,4\,6 \end{array}$$

Round:
$$\begin{array}{r} 7,0\,0\,0 \\ -1,0\,0\,0 \\ \hline 6,0\,0\,0 \end{array}$$

PRACTICE

Use front-end estimation for the problems below.

1
$$\begin{array}{r} \$25.91 \\ -\ 12.03 \\ \hline \end{array}$$

2
$$\begin{array}{r} \$512.14 \\ -\ 211.91 \\ \hline \end{array}$$

3 $25,921 - 3,412 = $ ___

4 $\$67.44 - \$39.00 = $ ___

5 $71,892 - 25,912 = $ ___

In this column, round each number to the highest possible place. Then find the difference of the rounded numbers.

6
$$\begin{array}{r} 4,192 \\ -\ 1,527 \\ \hline \end{array}$$

7
$$\begin{array}{r} 3,192 \\ -\ 1,890 \\ \hline \end{array}$$

8
$$\begin{array}{r} 92,705 \\ -\ 32,905 \\ \hline \end{array}$$

9
$$\begin{array}{r} \$35.98 \\ -\ 11.05 \\ \hline \end{array}$$

10
$$\begin{array}{r} 8,221 \\ -\ 997 \\ \hline \end{array}$$

11
$$\begin{array}{r} 497 \text{ feet} \\ -\ 97 \text{ feet} \\ \hline \end{array}$$

In this column, round each amount to the nearest dollar. Then find the difference of the rounded numbers.

12 $19.21, $13.42

13 $75.89, $38.79

14 $61.85, $39.99

15 $6.99, $2.97

Solving Word Problems

You can use the same five steps to solve any word problem. Review the five steps. Then use them to solve the problems below.

1.	Identify the particular question in the problem.
2.	Get all the information you need.
3.	Set up the problem.
4.	Solve it.
5.	Check your work.

PRACTICE

What does each question ask for? Circle the letter for the best answer. *Remember:* In a subtraction problem, you to take something away or find a difference. In an addition problem, you combine two amounts.

1   Zeke spent $12.89 on gas. He paid with a 20-dollar bill. About how much change should he get back?

   A   an exact sum
   B   an exact difference
   C   an estimated difference
   D   an estimated sum

2   Linda's hair is 12 inches long. If it grows 2 inches next month, how long will it be?

   F   an exact sum
   G   an exact difference
   H   an estimated difference
   J   an estimated sum

3   Martin bought 50 pounds of potatoes. There are only 15 pounds of potatoes left. How many pounds of potatoes have been used?

   A   an exact sum
   B   an exact difference
   C   an estimated difference
   D   an estimated sum

4   Jay is 25 years old and Kay is 37. How much older is Kay?

   F   an exact sum
   G   an exact difference
   H   an estimated difference
   J   an estimated sum

5   There are just over 210 days in the school year. This is the 175th day of school. About how many days of school are left?

   A   an exact sum
   B   an exact difference
   C   an estimated difference
   D   an estimated sum

6   Diego bought 2 crates of potato chips. Each crate contains 52 bags of chips. Approximately how many bags of chips did he buy?

   F   an exact sum
   G   an exact difference
   H   an estimated difference
   J   an estimated sum

7   Last month there was $1,542 in Johanna's bank account. This month the account balance is $1,837. How much money did Johanna save this month?

   A   an exact sum
   B   an exact difference
   C   an estimated difference
   D   an estimated sum

Use the information below to set up and solve problems 8–11. Then use estimation to check whether your answer is reasonable. Remember: You can only add and subtract like things.

Edwardo's company has 892 packages to deliver. He will deliver 381 of them to addresses in town. The rest are going to places in the country. Edwardo thinks that one driver working in town can deliver 35 packages a day. A driver working in the country can deliver about 15 packages a day.

8 How many packages must be delivered to the country?

9 How many more packages can a city driver deliver in a day than a country driver can deliver?

10 Edwardo's workers deliver 559 packages on the first day. How many packages are left?

11 In a day, one city driver and one country driver can deliver about how many packages?

These word problems do not give enough information. On the blank lines, tell what information is needed to solve each problem.

12 Eddie only has $5.10 left in his wallet. How much has he spent?

You also need to know _____

_____

13 Arlene has saved $253.00 to pay for a new dining room set. How much more does she need to save?

You also need to know _____

_____

14 Sandy bought 400 pounds of cement to build a patio. He will use what is left over to build a small fish pond. How much cement will he have left for the pond?

You also need to know _____

_____

15 Sylvia was caught in a traffic jam today. It took her 96 minutes to drive to work. How much longer than her usual drive was today's drive?

You also need to know _____

_____

Solving Word Problems

Subtraction Skills Practice

Circle the letter for the best answer to each problem. Try crossing out any unreasonable answers before you start to work on each problem.

1
$$395 - 142$$

  A 243
  B 252
  C 242
  D 253
  E None of these

2
$$2{,}342 - 331$$

  F 2,011
  G 2,042
  H 2,111
  J 2,311
  K None of these

3
$$507 - 63$$

  A 444
  B 544
  C 443
  D 545
  E None of these

4
$$5{,}419 - 170$$

  F 5,369
  G 5,349
  H 5,249
  J 5,449
  K None of these

5
$$2{,}450 - 14 = \underline{\qquad}$$

  A 2,310
  B 1,050
  C 2,446
  D 2,436
  E None of these

6
$$96 - 42 = \underline{\qquad}$$

  F 54
  G 45
  H 52
  J 53
  K None of these

7
$$211 - 35$$

  A 166
  B 186
  C 175
  D 176
  E None of these

Use the following information to do Numbers 8 through 10.

Trina was out of cash, so she took $50.00 out of the bank. Then she spent $12.50 on gasoline.

8 How can you figure out how much money Trina has left?

  F Add.
  G Divide.
  H Multiply.
  J Subtract.

9 Trina gives her daughter $3.50 for lunch. How much has Trina spent altogether?

  A $53.50
  B $15.50
  C $16.00
  D $66.00

10 Before she withdrew the money, Tina had $1,479.00 in her account. How much does she have in the account now?

  F $1,528.00
  G $1,474.00
  H $979.00
  J $1,429.00

11

$$\begin{array}{r} 88 \\ -\ 62 \end{array}$$

A 26
B 16
C 27
D 25
E None of these

12

What is
$12.00 − $8.00?

F $0.40
G $4.00
H $40.00
J $4.50
K None of these

13

$$\begin{array}{r} 5376 \\ -\ 1354 \end{array}$$

A 422
B 4,022
C 4,021
D 4,032
E None of these

14

What is
607 − 90?

F 517
G 617
H 697
J 597
K None of these

15

$$\begin{array}{r} \$10.90 \\ -\ 8.05 \end{array}$$

A $2.90
B $2.10
C $2.95
D $2.85
E None of these

16

$$\begin{array}{r} 6{,}850 \\ -\ 5{,}350 \end{array}$$

F 1,600
G 150
H 1,050
J 1,550
K None of these

17 Which of the following, if any, is false?

A $1{,}679 - 1{,}679 = 0$
B If $97 - 18 = 79$,
then $79 + 18 = 97$.
C $34 - 10 - 6 = 34 - 16$
D $34 - 10 - 6 = 34 - 4$

18 What number goes in the box to make the second number sentence true?

$345 + 52 = 397$
$397 - \boxed{} = 345$

F 52
G 397
H 342
J It is impossible to tell.

19 Which of these, if any, has the same value as $37 - 9$?

A $9 - 37$
B $37 - 9 - 0$
C $37 - 9 - 1$
D None of these

20 Which of the following phrases represents "$89 - 25$" ?

F twenty-five less than eighty-nine
G eighty-nine divided by twenty-five
H eighty-nine subtracted from twenty-five
J eighty-nine more than twenty-five

Multiplication

Basic Concepts

You can think of multiplication as a shortcut for adding the same number over and over again.

Addition	Multiplication
9	
9	9
9	$\times\ 4$
$+\ 9$	3 6
3 6	Think: 4 times 9 is 36.

- The symbol × is a multiplication symbol. The answer in a multiplication problem is called the **product.**

- If you change the order of the numbers being multiplied, you do not change the product.

  3×4 and 4×3 have the same value.

- If you multiply any number times zero, the product is zero.

  $5 \times 0 = 0$ $0 \times 23 = 0$

- To find the product of several numbers, it does not matter which pair of numbers you start with.

  $$2 \times 3 \times 3 = (2 \times 3) \times 3 = 6 \times 3 = 18$$
  $$= 2 \times (3 \times 3) = 2 \times 9 = 18$$

- If you multiply a number by 1, you do not change the value.

  $3,471 \times 1 = 3,471$

PRACTICE

Fill in the blanks.

1   $4,562 \times 1 = $ _____

2   $72,697 \times 0 = $ _____

3   $65 + 65 + 65 + 65 = $ _____
(Write a multiplication problem.)

4   $59 \times 3 = $ _____
(Write an addition problem.)

5   $4 \times 3 \times 2 = $ _____ × _____

     $= $ _____ × _____
(Write the product two different ways.)

Write T for *true* or F for *false*.

6   17×3 has the same value as 3×17. _____

7   $12 \times 4 = 4 + 4 + 4 + 4$ _____

8   $3 \times 12 = 12 + 12 + 12$ _____

9   $322 \times 0 = 322$ _____

10   $4,103 \times 1 = 1$ _____

11   $165 \times 0 = 0$ _____

12   $8 \times 2 \times 3$ has the same value as 8×6.   _____

Basic Multiplication Facts

To solve problems using multiplication, you must know the basic facts quickly and accurately. If you make more than a few mistakes on these problems, erase your answers and try the problems again.

PRACTICE

Remember: You can change the order of the numbers you are multiplying without changing the product. So if you know that
$$9 \times 3 = 27$$
you also know that
$$3 \times 9 = 27$$

1 $2 \times 6 =$ _____ $4 \times 4 =$ _____

2 $2 \times 3 =$ _____ $8 \times 4 =$ _____ $6 \times 6 =$ _____

3 $6 \times 5 =$ _____ $5 \times 3 =$ _____ $7 \times 6 =$ _____

4 $4 \times 2 =$ _____ $7 \times 4 =$ _____ $8 \times 9 =$ _____

5
$$\begin{array}{r} 4 \\ \times 3 \\ \hline \end{array} \qquad \begin{array}{r} 2 \\ \times 2 \\ \hline \end{array} \qquad \begin{array}{r} 7 \\ \times 3 \\ \hline \end{array} \qquad \begin{array}{r} 3 \\ \times 9 \\ \hline \end{array}$$

6
$$\begin{array}{r} 7 \\ \times 7 \\ \hline \end{array} \qquad \begin{array}{r} 8 \\ \times 3 \\ \hline \end{array} \qquad \begin{array}{r} 3 \\ \times 3 \\ \hline \end{array} \qquad \begin{array}{r} 4 \\ \times 5 \\ \hline \end{array}$$

7
$$\begin{array}{r} 5 \\ \times 9 \\ \hline \end{array} \qquad \begin{array}{r} 7 \\ \times 5 \\ \hline \end{array} \qquad \begin{array}{r} 2 \\ \times 7 \\ \hline \end{array} \qquad \begin{array}{r} 4 \\ \times 6 \\ \hline \end{array}$$

8
$$\begin{array}{r} 8 \\ \times 2 \\ \hline \end{array} \qquad \begin{array}{r} 2 \\ \times 9 \\ \hline \end{array} \qquad \begin{array}{r} 8 \\ \times 5 \\ \hline \end{array} \qquad \begin{array}{r} 8 \\ \times 8 \\ \hline \end{array}$$

9
$$\begin{array}{r} 9 \\ \times 4 \\ \hline \end{array} \qquad \begin{array}{r} 7 \\ \times 8 \\ \hline \end{array} \qquad \begin{array}{r} 5 \\ \times 5 \\ \hline \end{array} \qquad \begin{array}{r} 6 \\ \times 8 \\ \hline \end{array}$$

10
$$\begin{array}{r} 7 \\ \times 9 \\ \hline \end{array} \qquad \begin{array}{r} 3 \\ \times 6 \\ \hline \end{array} \qquad \begin{array}{r} 9 \\ \times 9 \\ \hline \end{array} \qquad \begin{array}{r} 6 \\ \times 9 \\ \hline \end{array}$$

Multiplying by a One-Digit Number

Here is a multiplication problem in column form. Working from right to left, multiply each digit in the top number by the bottom number.

$$
\begin{array}{r}
3\,4 \\
\times\ 2 \\
\hline
6\,8
\end{array}
$$

Multiply by this digit.

$2 \times 4 = 8$
$2 \times 3 = 6$

PRACTICE

Find each product. Then review your work to make sure you have done it correctly. *Hint:* **Write every digit in a product, even if it is zero.**

> If one of the numbers in a problem has a label, repeat it in your answer. Also, remember that dollar amounts always have two digits to the right of the decimal point.

1
$$
\begin{array}{r} 24 \\ \times 2 \\ \hline \end{array}
\qquad
\begin{array}{r} 232 \\ \times\ 3 \\ \hline \end{array}
\qquad
\begin{array}{r} \$1.22 \\ \times\ \ 4 \\ \hline \end{array}
\qquad
\begin{array}{r} 31 \\ \times 5 \\ \hline \end{array}
\qquad
\begin{array}{r} 41 \text{ inches} \\ \times 6 \\ \hline \end{array}
$$

2
$$
\begin{array}{r} 204 \\ \times\ 2 \\ \hline \end{array}
\qquad
\begin{array}{r} 81 \text{ feet} \\ \times 5 \\ \hline \end{array}
\qquad
\begin{array}{r} \$1.00 \\ \times\ \ 9 \\ \hline \end{array}
\qquad
\begin{array}{r} 74 \\ \times 2 \\ \hline \end{array}
\qquad
\begin{array}{r} 92 \\ \times 3 \\ \hline \end{array}
$$

Rewrite each problem in column form. Then solve the problem. Review your work to make sure it is correct.

3 $214 \times 2 =$ _____

4 $3 \times 43 =$ _____

5 $5 \times 70 =$ _____

6 $810 \times 5 =$ _____

7 $33 \times 3 =$ _____

8 $\$4.13 \times 2 =$ _____

9 $\$9.44 \times 2 =$ _____

10 $7{,}003 \times 3 =$ _____

11 $\$8.00 \times 8 =$ _____

Multiplying by a Two-Digit Number

In the problem below, you are multiplying by the two-digit number 23.

◆ First, multiply the top number by the ones digit of the bottom number.
◆ Next, multiply the top number by the tens digit of the bottom number. Start your answer in the tens column, leaving a blank space in the ones place.
◆ Finish the problem by adding the two products together.

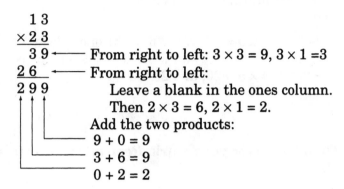

```
      1 3
    × 2 3
      3 9  ◄——— From right to left: 3 × 3 = 9, 3 × 1 = 3
    2 6    ◄——— From right to left:
    2 9 9           Leave a blank in the ones column.
                    Then 2 × 3 = 6, 2 × 1 = 2.
                 Add the two products:
                    9 + 0 = 9
                    3 + 6 = 9
                    0 + 2 = 2
```

PRACTICE

Find each product. In the first two problems, some of the work has been done.

1
```
    11        121        302        21        43
  × 13      × 22       × 13      × 25      × 21
    33        242
    11        242
```

2
```
   400       610         80        63        95
  × 52      × 15       × 36      × 31      × 11
```

Rewrite these problems in column form. Then solve them.

3 231 × 30 = _____

4 513 × 32 = _____

5 30 × 71 = _____

6 302 × 23 = _____

7 44 × 22 = _____

8 50 × 51 = _____

9 900 × 22 = _____

10 200 × 16 = _____

11 860 × 11 = _____

When a Column Product Is More Than 9

In the multiplication problem below, the product in the ones column is a two-digit number. You write the digit 6 in the ones place as part of your answer, and you write the digit 3 above the tens place. After you find the product 4 × 5, you add the 3.

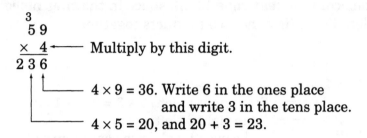

Multiply by this digit.

4 × 9 = 36. Write 6 in the ones place and write 3 in the tens place.

4 × 5 = 20, and 20 + 3 = 23.

PRACTICE

Solve each problem. Then look over your work. *Remember:* **Multiply** *before* **you add the digit being carried.**

1
$$\begin{array}{r} 32 \\ \times 5 \\ \hline \end{array}$$

2
$$\begin{array}{r} 65 \\ \times 4 \\ \hline \end{array}$$

3
$$\begin{array}{r} 255 \\ \times\ 6 \\ \hline \end{array}$$

4
$$\begin{array}{r} 39 \\ \times 5 \\ \hline \end{array}$$

5
$$\begin{array}{r} 32 \\ \times 7 \\ \hline \end{array}$$

6
$$\begin{array}{r} 125 \\ \times\ 5 \\ \hline \end{array}$$

7 308 × 4 = _____

8 28 × 5 = _____

9 95 × 6 = _____

10 709 × 7 = _____

11 25 × 50 = _____

12
$$\begin{array}{r} 44 \\ \times 15 \\ \hline \end{array}$$

Hint: When you are multiplying by a 2-digit number, erase all the carry digits after your first multiplication. That gives you room to write carry digits for the second multiplication.

13
$$\begin{array}{r} 252 \\ \times 34 \\ \hline \end{array}$$

14
$$\begin{array}{r} 731 \\ \times\ 5 \\ \hline \end{array}$$

15
$$\begin{array}{r} 19 \\ \times 43 \\ \hline \end{array}$$

Using Estimation When You Multiply

To estimate a product, round each number to its highest place value and then multiply. However, you do not round a one-digit number.

Problem 1	Rounded	Problem 2	Rounded
3,784	4,000	894	900
× 9	× 9	× 19	× 20
	36,000		000
			1800
			18,000

PRACTICE

Use rounding to estimate the answer to each problem. *Hint:* **Your estimate should have at least one zero for every zero in the rounded numbers.**

Shortcut: You can work even faster if you
1. Round each number to its highest place value.
2. Multiply the front-end digits.
3. Write one zero for every zero in the rounded numbers.

1   67
    × 12

2   97
    × 29

3   1,927
    × 9

4 $29.00
    × 3

5 $12.59
    × 5

6   87 feet
    × 18

7 $129
    × 5

8   459
    × 72

9   78 × 81 = _____

10   603 × 92 = _____

11   789 × 63 = _____

12   92 × 37 = _____

Solving Word Problems

An important part of solving any word problem is deciding whether you should add, multiply, subtract, or divide (you will explore division in the next section). Look for these clues.

- If you put different amounts together, you will add.
- If you compare one amount to another, you will subtract or divide.
- If you take something away from something else, you will subtract.
- If there are several of something, you will multiply or divide.

Signal Words

Addition	Subtraction	Multiplication
plus	minus	times
sum	difference	multiplied by
total	take away	product
added to	subtract	twice, three times, and so on
altogether	left over	two or more of something
in all	how much change	apiece
combined	how much more than	each
increased by	how much less than	
	decreased by	

PRACTICE

For each problem, tell whether the best method is to add, subtract, or multiply. You do not have to solve the problem.

1   There are 23 tables in the cafe. There are 4 seats at each table. How many seats are there in the cafe?

    A Add.
    B Subtract.
    C Multiply.

2   Amanda is covering one of her kitchen walls with square mirrors. There will be 12 rows of mirrors. Each row will contain 24 mirrors. How many mirrors will there be in all?

    F Add.
    G Subtract.
    H Multiply.

3   Katherine bought 12 shirts for her family. She got each shirt for $12.00 off. How much money did she save?

    F Add.
    G Subtract.
    H Multiply.

4   Sangeeta is collecting toys for charity. She picked up 27 boxes of toys at the local high school. The Benefactor Club gave her another 25 boxes of toys. How many boxes does she have?

    A Add.
    B Subtract.
    C Multiply.

Julius is a telemarketer. He makes $4.00 on every sale he makes. This list shows how many sales he made last week.

Monday	32
Tuesday	27
Wednesday	10
Thursday	30
Friday	6

Use this information to set up and solve problems 5–8. Use estimation to check your work.

5 How much money did Julius make on Monday?

7 How many sales did Julius make in all last week?

6 How many more sales did Julius make on Monday than he made on Friday?

8 On average, Julius made 21 sales a day. At that rate, about how many sales will he make in 22 work days? (Estimate.)

These word problems do not give enough information. On the blank lines, tell what else you would need to know before you could solve each problem.

9 Rent City charges $21 a day to rent a small car, $29 a day for a mid-size car, and $42 a day for a sedan. Eric rented a car for 3 days. How much did he owe?

You also need to know _____

_____

10 The caterer at Jeanette's wedding charged $19.00 per person. What was the total charge?

You also need to know _____

_____

11 Carlos is building a brick patio. There will be 2 bricks per square foot of patio. How many bricks does Carlos need?

You also need to know _____

_____

Multiplication Skills Practice

Circle the letter for the best answer to each problem. Try crossing out unreasonable answers before you start to work on each problem.

1.
$$\begin{array}{r} 10 \\ \times\,9 \\ \hline \end{array}$$

  A 19
  B 90
  C 99
  D 900
  E None of these

2.
$$\begin{array}{r} 78 \\ \times\,3 \\ \hline \end{array}$$

  F 224
  G 234
  H 254
  J 2,324
  K None of these

3.
$521 \times 24 = \underline{\hspace{1cm}}$

  A 13,026
  B 13,126
  C 12,504
  D 12,404
  E None of these

4.
$$\begin{array}{r} 39 \\ \times\,5 \\ \hline \end{array}$$

  F 195
  G 155
  H 185
  J 355
  K None of these

5.
$$\begin{array}{r} 150 \\ \times\,3 \\ \hline \end{array}$$

  A 45
  B 35
  C 350
  D 650
  E None of these

6.
$$\begin{array}{r} 240 \\ \times\,32 \\ \hline \end{array}$$

  F 120
  G 768
  H 668
  J 7,248
  K None of these

Marco is having a barbecue for 15 people. He figures that he needs the following supplies for each person who comes. **Study the list. Then do Numbers 7 through 9.**

> 3 cans soda
> 2 bags potato chips
> 1 pound meat
> 2 hamburger buns
> 4 ounces baked beans
> 1 cup potato salad

7. How many cans of soda should Marco buy?

  A 18
  B 30
  C 45
  D 35

8. Steak costs $5.00 a pound. If Marco buys steak for the barbecue, how much would it cost?

  F $20.00
  G $55.00
  H $7.50
  J $75.00

9. Marco can buy German potato salad for $15.00. He can buy regular potato salad for $13.50. To find how much more the German potato salad costs than the regular potato salad, what should Marco do?

  A Add.
  B Subtract.
  C Multiply.
  D Divide.

10

$61 \times 10 =$ _____

    F   71
    G   610
    H   6,100
    J   61
    K   None of these

11

    203
    $\times\ 4$

    A   812
    B   8,012
    C   612
    D   807
    E   None of these

12

$\$21.00 \times 2 =$ _____

    F   \$23.00
    G   \$420.00
    H   \$41.00
    J   \$42.00
    K   None of these

13

    243
    $\times\ 12$

    A   2,918
    B   2,816
    C   2,916
    D   729
    E   None of these

14

$1,850 \times 6 =$ _____

    F   10,830
    G   10,800
    H   1,110
    J   11,100
    K   None of these

15

    92
    $\times 6$

    A   542
    B   552
    C   642
    D   462
    E   None of these

16

$32 \times 43 =$ _____

    F   1,376
    G   1,276
    H   224
    J   1,346
    K   None of these

17 Which of the following, if any, equals 57 ?

    A 57×0
    B 57×1
    C 57×57
    D   None of these

18 Which of the following problems, if any, can be solved with multiplication?

    F   Larry has \$2,000 in his vacation fund. He spends \$559 on plane tickets. How much is left?
    G   Clint buys a shirt for \$52.00, a hat for \$62.75, and shoes for \$89.00. How much did he spend?
    H   Jorge buys 24 pears for 51 cents each. How much did he spend?
    J   None of these

19 Which of these, if any, has the same value as 41×3 ?

    A $41 + 41 + 41$
    B $41 \times 2 \times 1$
    C $41 \times 3 \times 0$
    D   None of these

20 Which of these *does not* have the same value as 67×20 ?

    F 20×67
    G $67 \times 20 \times 1$
    H $67 \times 4 \times 5$
    J $67 \times 20 \times 0$

Division

Basic Concepts

The operation of separating a group of objects or a number into smaller groups or smaller numbers is called **division**.

Example 1:
You must divide $84.00 among 7 people. How much does each person get?

Example 2:
Lottery tickets cost $2.50. How many can you buy for $15.00?

- Two symbols for division are $\div$ and $\overline{)}$. The answer in a division problem is called the **quotient**.

Division problems: $990 \div 3 =$ ____ $3\overline{)990}$

- If you divide a number by 1, you do not change its value. If you divide a number by itself, the quotient is 1.

$92 \div 1 = 92$

$92 \div 92 = 1$

- The order of the numbers in a division problem is important.

$12 \div 4$ and $4\overline{)12}$ are the same division problem. They are different from $4 \div 12$ and $12\overline{)4}$.

- If you divide zero by any (nonzero) number, the result is zero. You cannot divide any number by zero.

$0 \div 71 = 0$

$71 \div 0$ does not have any meaning in math.

PRACTICE

1   **Circle the letter for the word problems that call for division.**

   A   Cherry paid $1,200 for a 5-day cruise. How much did she pay per day?
   B   Cherry had $2,560.23 in her bank account before she paid for the trip. How much did she have afterwards?
   C   On the trip, Cherry spent $63.30 on gifts for her sisters and $38.92 on a dress for herself. How much did she spend in all?
   D   If Cherry saves $75 per month, how many months will it take to put the $1,200.00 back in her bank account?

Fill in the blanks.

2   $671 \div 1 =$ ____

3   $0 \div 89 =$ ____

4   $75 \div 5 = 25.$
   Does $5 \div 75 = 25?$ ____

5   $367 \div 367 =$ ____

6   $66 \div 1 =$ ____

Basic Concepts

Basic Division Facts

Each basic division fact is related to a multiplication fact. For example, a multiplication fact is $6 \times 7 = 42$. Two related division facts are $42 \div 7 = 6$ and $42 \div 6 = 7$.

You should know the division facts quickly and accurately. Do the problems below. If you make more than a few mistakes, erase your answers and try them again.

PRACTICE

1	$12 \div 6 = $ _____	$16 \div 4 = $ _____	$8 \div 4 = $ _____	$18 \div 6 = $ _____
2	$6 \div 3 = $ _____	$32 \div 4 = $ _____	$36 \div 6 = $ _____	$20 \div 4 = $ _____
3	$30 \div 5 = $ _____	$15 \div 3 = $ _____	$9 \div 3 = $ _____	$8 \div 2 = $ _____
4	$10 \div 2 = $ _____	$28 \div 4 = $ _____	$10 \div 5 = $ _____	$6 \div 2 = $ _____
5	$12 \div 3 = $ _____	$4 \div 2 = $ _____	$21 \div 3 = $ _____	$27 \div 3 = $ _____
6	$49 \div 7 = $ _____	$24 \div 3 = $ _____	$42 \div 6 = $ _____	$35 \div 5 = $ _____
7	$45 \div 5 = $ _____	$35 \div 7 = $ _____	$14 \div 7 = $ _____	$24 \div 6 = $ _____
8	$16 \div 2 = $ _____	$18 \div 2 = $ _____	$40 \div 5 = $ _____	$64 \div 8 = $ _____
9	$36 \div 4 = $ _____	$56 \div 8 = $ _____	$25 \div 5 = $ _____	$48 \div 8 = $ _____
10	$63 \div 9 = $ _____	$18 \div 3 = $ _____	$81 \div 9 = $ _____	$54 \div 9 = $ _____
11	$12 \div 2 = $ _____	$24 \div 4 = $ _____	$15 \div 5 = $ _____	$20 \div 5 = $ _____
12	$48 \div 6 = $ _____	$42 \div 7 = $ _____	$54 \div 6 = $ _____	$28 \div 7 = $ _____

Using the Division Bracket $\overline{)}$

To write a division problem using the bracket $\overline{)}$, the number you are *dividing up* (the **dividend**) goes inside the bracket. The number you are *dividing by* (the **divisor**) goes to the left of the bracket.

Write $105 \div 5$ as $5\overline{)105}$.
Then $5\overline{)105}^{\,21}$

PRACTICE

Rewrite each problem using the division bracket. You do not have to solve the problems.

1 $70 \div 5 =$ _____

2 $98 \div 2 =$ _____

3 $24 \div 12 =$ _____

4 $142 \div 12 =$ _____

5 What is 150 divided by 10?

6 What is 575 divided by 25?

7 What is ninety divided by fifteen?

8 What is eighty-one divided by nine?

9 What is 212 inches divided by 4?

10 What is 98 feet divided by 2?

11 You must divide 16 tokens among 4 people. How many tokens will each person get?

12 After a meal, Lauren offered to pay the bill. She paid $20.00 for 5 hamburgers. How much did each hamburger cost?

13 Giles has a concert 1,200 miles away. He has three days to get there. If he divides the trip up evenly, how many miles will he drive each day?

14 Rent is $654 per month. Three roommates share the cost evenly. How much rent does each roommate pay per month?

15 You used 12 gallons of gas during a 250-mile trip. How far did you drive on each gallon of gas?

16 Helena must knit 12 pairs of mittens before Christmas. Christmas is 24 days away. On average, how many days does she have to knit each pair of mittens?

Dividing by a One-Digit Number

In this problem, the number you are *dividing by* (the divisor) has one digit. Start at the left of the dividend 48. First, divide 2 into 4. Then divide 2 into 8.

Divide 2 into 4: $4 \div 2 = 2$.
Divide 2 into 8: $8 \div 2 = 4$.

$$2\,4 \leftarrow \text{The quotient is 24.}$$
$$2\overline{)4\,8}$$

Since multiplication and division are the opposite of each other, each will "undo" the other. You can use this to check a division problem. Multiply the quotient 24 times the divisor 2. The result should be the dividend 48.

$$
\begin{array}{r}
2\,4 \leftarrow \text{quotient} \\
\times\ \ 2 \leftarrow \text{divisor} \\
\hline
4\,8
\end{array}
$$

PRACTICE

Find each quotient. Then use multiplication to check your work. *Remember:* **Zero divided by any number is zero.**

1   $4\overline{)84}$

2   $3\overline{)96}$

3   $2\overline{)268}$

4   $3\overline{)336}$

5   $2\overline{)824}$

6   $7\overline{)707}$

7   $8\overline{)88}$ inches

8   $2\overline{)608}$ dollars

9   $3\overline{)963}$ inches

Rewrite each problem using the division symbol $\overline{)}$. Then solve the problem and check your work. *Remember:* **The number you are *dividing up* goes inside the bracket.**

10   What is 844 divided by 4?

11   $5,055 \div 5 = $ _____

12   Jorge has 26 dollars. Carnival tickets cost 2 dollars each. How many tickets can Jorge buy?

13   Virginia has 84 duplicate family photographs. If she divides them evenly among her four children, how many photos will each child get?

Dividing by a One-Digit Number

When a Dividend Digit Is Too Small

In this problem, the 4 inside the division symbol is smaller than the divisor 6. To start, divide 6 into the first two digits of 426.

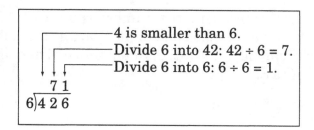

PRACTICE

Solve each problem. Then check your work. Divide into every digit under the division bracket, even if that digit is a zero.

1 9)189

2 7)147

3 8)2400

4 155 divided by 5

5 5)405

6 4)1680

7 204 divided by 4

8 6)186 inches

9 6)366 pounds

After you write the first digit in the quotient, use a zero as a placeholder whenever a digit in the dividend (inside the bracket) is smaller than the divisor.

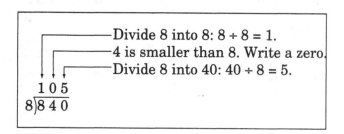

10 7)735

11 9)981

12 5)545

13 6)624

14 9,021 divided by 3

15 2)812

16 4)812

17 5)1045

18 749 tons divided by 7

19 3)2109

20 2)218

21 9)927

Dividing with a Remainder

One number cannot always be evenly divided into another. For instance, 9 objects cannot be divided into 2 equal piles. However, 9 objects can be divided into 2 piles of 4 objects each, with 1 extra object left over. Therefore we say that "9 divided by 2 is 4, with a remainder of 1."

$$9 \div 2 = 4 \text{ r } 1$$

To solve a problem with a remainder, you must find the largest whole number that goes into the dividend digit(s). The next step is to multiply that digit by the divisor. The last step is to subtract.

$$
\begin{array}{r}
3 \\
8\overline{)29} \\
24 \\
\hline
5
\end{array}
$$

— $29 \div 8$ is a little more then 3. Write 3.

— Multiply: $3 \times 8 = 24$.

— Subtract: $29 - 24 = 5$.

The quotient is 3 r 5.

PRACTICE

Solve each problem. *Hint:* **The correct remainder is always smaller than the divisor.**

1 $7\overline{)8}$

2 $4\overline{)15}$

3 $7\overline{)30}$

4 $9\overline{)39}$

5 $9\overline{)19}$

6 $8\overline{)19}$

7 $6\overline{)13}$

8 $8 \div 3 =$ ____

9 $11 \div 2 =$ ____

10 $14 \div 5 =$ ____

11 $14 \div 6 =$ ____

12 What is 18 divided by 5?

13 What is 91 divided by 9?

To check a problem with a remainder such as the sample problem above:
1. Multiply: $3 \times 8 = 24$.
2. Add the remainder: $24 + 5 = 29$.
3. The result should be the original dividend.

14 Fifty-two cards must be divided among five people. How many cards will each person get, and how many cards will be left over?

15 Otis buys 24 donuts for his work crew. There are 7 workers on the crew. How many donuts are there for each worker, and how many donuts are left over?

Dividing with a Remainder

Writing the Steps of a Division Problem

In the problem below, the divisor 6 does not evenly divide the first two digits of the dividend 474.

Here is a step-by-step method for solving such a problem.

```
      7 9
6 ) 4 7 4      1. Divide 6 into 47. Write 7 in the quotient.
    4 2         2. Multiply: 7 × 6 = 42. Write 42.
    ___
      5 4       3. Subtract: 47 − 42 = 5. Bring down the next digit 4.
      5 4       4. Divide 6 into 54: 54 ÷ 6 = 9. Write 9 in the quotient.
      ___
        0       5. Multiply: 9 × 6 = 54. Subtract: 54 − 54 = 0.
```

The quotient is 79, with no remainder.

PRACTICE

Solve each problem. Check your work by multiplying the quotient times the divisor. Make sure you write down every step, including the subtraction.

1 3)75

2 2)58

3 5)65

4 3)81

5 7)84

6 5)70

7 6)84

8 8)968

9 6)672

10 575 ÷ 5 = _____

11 917 ÷ 7 = _____

12 7)4340

13 What is 324 divided by 3?

14 What is 905 divided by 5?

15 What is 42 divided by 3?

16 An 144-foot stretch of land must be marked off in 6-foot sections. How many sections will there be?

Long Division

Look at the problem at the right and the problem below. Both problems show the general procedure for the steps of **long division.**

```
     5 8 r 3
4 ) 2 3 5
    2 0
    ---
      3 5
      3 2
      ---
        3
```

1. Divide 4 into 23. Write 5 in the quotient.
2. Multiply: $5 \times 4 = 20$. Write 20.
3. Subtract: $23 - 20 = 3$. Bring down the next digit 5.
4. Divide 4 into 35. Write 8 in the quotient and multiply: $8 \times 4 = 32$.
5. Subtract: $35 - 32 = 3$.

The quotient is 58 r 3.

Notice the pattern: after the first three steps, you always bring down the next digit, then divide, then multiply and then subtract.

```
       3 4 5
4 ) 1 3 8 0
    1 2
    ---
      1 8
      1 6
      ---
        2 0
        2 0
        ---
          0
```

1. Divide 4 into 13. Write 3.
2. Multiply: $3 \times 4 = 12$.
3. Subtract: $13 - 12 = 1$.
4. Bring down the next digit 8.
5. Divide 4 into 18. Write 4.
6. Multiply: $4 \times 4 = 16$.
7. Subtract: $18 - 16 = 2$.
8. Bring down the next digit 0.
9. Divide 4 into 20. Write 5.
10. Multiply: $5 \times 4 = 20$.
11. Subtract: $20 - 20 = 0$.

PRACTICE

Use long division to solve these problems. Show all your work, and check your answers using multiplication.

1 $4 \overline{)540}$

2 $8 \overline{)176}$

3 $5 \overline{)720}$

4 $9 \overline{)198}$

5 $3 \overline{)534}$

6 $2 \overline{)330}$

Dividing by a Two-Digit Number

When you divide by a two-digit number, start by finding the first digit for the quotient. Look at these two examples.

Example 1

```
      2 1
15)3 1 5
   3 0
     1 5
     1 5
        0
```

1. Divide 31 by 15. Write 2.
2. Multiply: $2 \times 15 = 30$.
3. Subtract: $31 - 30 = 1$.
4. Bring down the next digit 5.
5. Divide 15 by 15. Write 1.
6. Multiply: $1 \times 15 = 15$.
7. Subtract: $15 - 15 = 0$.

Example 2

```
       1 3
23)2 9 9
     2 3   ←── 1 × 23
       6 9
       6 9 ←── 3 × 23
          0
```

Important: The results of the "multiply" step *must not* be larger than the number above it. If it *is* larger, do the step again using a smaller digit in the divisor.

PRACTICE

Solve each problem below. Then check your work. *Hint:* **These problems do not have remainders.**

1 15)30

2 22)66

3 11)121

4 21)651

5 25)50

6 15)45

7 45)90

8 12)264

9 26)78

Using Estimation When You Divide

One way to estimate a quotient is to find the first digit in the answer. Then add a zero in the quotient for each of the other digits in the dividend.

The exact answer will always be **more than** your estimate.

$$\begin{array}{r} 4\,0\,0 \\ 5\overline{)2\,1\,4\,3} \end{array}$$

The first digit in the quotient is 4.

Write a zero in the quotient for each of the other digits in the dividend.

PRACTICE

Estimate the quotient for each problem. Check to make sure you have added the right number of zeros.

1 $3\overline{)1267}$

2 $6\overline{)364}$

3 $7\overline{)816}$

4 $4\overline{)513}$

5 $4\overline{)1091}$

6 $9\overline{)1132}$

7 $33\overline{)982}$

8 $25\overline{)564}$

9 $8\overline{)9320}$

10 $40\overline{)890}$

Circle the letter for the best estimate for each problem.

11 $3\overline{)628}$
A 100
B 20
C 200
D 10

12 $19\overline{)203}$
F 100
G 10
H 200
J 20

13 $9\overline{)7218}$
A 1,000
B 800
C 700
D 500

14 $14\overline{)3033}$
F 100
G less than 100
H 200
J more than 200

15 $8\overline{)806}$
A 100
B 10
C more than 100
D less than 100

Solving Word Problems

Set up each problem and then solve it. Not all of these are division problems. If you need to review signal words, see page 38.

Luisa is running the art booth at the Children's Fair. This year the children are painting T-shirts. Luisa has 250 T-shirts and 5 boxes of fabric paint. **Use this information to answer questions 1 through 4.**

1 Each box contains 25 tubes of fabric paint. How many tubes of paint does Luisa have in all?

2 Luisa paid 60 dollars for the fabric paint. If each box cost the same, what was the cost of each box of paint?

3 After one hour, 45 T-shirts have been used. At this rate, how many T-shirts will be used in 4 hours?

4 At the end of the fair, 27 T-shirts are left. How many T-shirts were used?

This list shows the Sakurada family's utility bills for March.

Electricity	$93.00
Telephone	$51.00
Cooking Gas	$14.00
Cable	$30.00

Use this information to answer questions 5 through 8.

5 How much higher was the electricity bill than the telephone bill?

6 There are 31 days in March. About how much did the family pay per day for cable during the month?

7 How much did the family pay in all for utilities?

8 If the family canceled its account for cable, how much would they save per year?

Solving Two-Step Word Problems

For each problem below, circle the letter for the choice that correctly describes how to solve the problem.

1 Darren has 3 six-packs of soda, plus 2 free cans. **How many cans of soda does he have in all?**

   A Add 3 and 2. Then multiply the sum by 6.
   B Multiply 3 by 6. Then add 2.
   C Multiply 3 by 6. Then subtract 2.

2 Bea gets three 20-dollar bills from the ATM machine. Then she spends $12.00. **How much of the ATM money does she have left?**

   F Subtract $12.00 from $20.00. Then multiply by 3.
   G Multiply $20.00 by 3. Then subtract $12.00.
   H Subtract $12.00 from $20.00.

3 Clarice buys 15 candy bars. She gives 3 of them to her husband. She saves the rest to be shared by her 4 children. **How many candy bars does she have for each child?**

   A Add 3 to 15. Then divide by 4.
   B Subtract 4 from 15. Then divide by 3.
   C Subtract 3 from 15. Then divide by 4.

4 Howard must drive 350 miles for a meeting. He will be able to drive 70 miles an hour all the way, but he plans to stop for an hour to have lunch with a friend. **How long will the trip take?**

   F Add 350 and 70. Then add 1.
   G Multiply 350 by 70. Then subtract 1.
   H Divide 350 by 70, then add 1.

Solve each two-step problem below.

5 Tickets to the fashion show are $18.00. The luncheon afterward costs $12.00 per person. **How much would it cost to take 3 people to the fashion show and the luncheon?**

6 Chris earns $500 repairing a porch. He spends $200 on materials and $10 on gas. **How much profit does he make?**

7 Michelle finds $3.00 in change in her pocket and $72.00 in her purse. Then she spends $50.00 at the mall. **How much money does she have left?**

8 The TV repair shop charges $40.00 per hour plus the cost of materials. It takes 2 hours to repair a TV, and the materials cost $15.50. **What will the bill be?**

Division Skills Practice

Circle the letter for the best answer to each problem. Try crossing out unreasonable answers before you start to work on each problem.

1
7)294

A 40 r 4
B 42
C 40 r 2
D 32
E None of these

2
6000 ÷ 4 = ____

F   150
G   1,200
H   1,400
J   1,500
K   None of these

3
10 ÷ 8 = ____

A   1
B   2
C   1 r 2
D   2 r 1
E   None of these

4
16 ÷ 3 = ____

F   5
G   6 r 1
H   6
J   5 r 1
K   None of these

5
3)9021

A   37
B   307
C   3,007
D   3,070
E   None of these

6
4)96

F   24
G   22 r 1
H   14
J   44 r 1
K   None of these

Use the following information to do Numbers 7 through 10.

Joe is having programs printed for the community play. He buys 600 programs for $50.00.

7   How many programs did Joe get for every dollar he spent?

A   12
B   83
C   60
D   50

8   There will be 3 performances of the play. At the first performance, 250 people saw the play. At this rate, how many will people will attend the play altogether?

F   300
G   650
H   750
J   7500

9   The programs are packaged in fifty boxes. What should Joe do to find out how many programs are in each box?

A   Add.
B   Subtract.
C   Multiply.
D   Divide.

10   Play tickets are $6.50 each. What is the correct change from a 10-dollar bill?

F $3.50
G $4.50
H $2.50
J $4.00

11

$4\overline{)220}$

A 55
B 50 r 2
C 550
D 44
E None of these

12

$4\overline{)8064}$

F 216
G 2,018
H 2,016
J 2,028
K None of these

13

$63 \div 9 = \underline{\hphantom{xxx}}$

A 9
B 8
C 7
D 6
E None of these

14

$97 \div 10 = \underline{\hphantom{xxx}}$

F 97
G 9 R 7
H 90 R 7
J 91
K None of these

15

$33 \div 3 = \underline{\hphantom{xxx}}$

A 10
B 11
C 12
D 10 r 3
E None of these

16

$306 \div 3 = \underline{\hphantom{xxx}}$

F 12
G 102
H 101
J 192
K None of these

17

$5\overline{)8}$

A 1
B 2
C 1 r 2
D 1 r 3
E None of these

18 Which of these, if any, has the same value as $95 \div 5$?

F $5 \div 95$
G $95\overline{)5}$
H $5\overline{)95}$
J None of these

19 What sign goes in the box to make the second number sentence true?

$78 \div 2 = 39$
$39 \ \boxed{} \ 2 = 78$

A ÷
B ×
C +
D –

20 Which of these number sentences is false?

F $84 \div 1 = 84$
G $57 \div 57 = 1$
H $0 \div 12 = 0$
J $1 \div 22 = 22$

Decimals

Decimal Place Values

In a number such as 2.63, the period is called a **decimal point** and the first three decimal place values to the right of the decimal point are **tenths, hundredths,** and **thousandths.** (Notice that each decimal place-value name ends in *-ths*.)

The number 2.63 is called a **decimal fraction** or, more simply, a **decimal.**

ones	tenths	hundredths	thousandths
2 .	6	3	__

Two and sixty-three hundredths

Amounts of money use decimals. A cent is a special name for *one hundredth of a dollar.* So another way to say $9.42 is "nine whole dollars and 42 hundredths of a dollar."

The 4 in $9.42 refers to tenths of a dollar.
The value of the 4 is $\frac{4}{10}$.

The 2 in $9.42 refers to hundredths of a dollar.
The value of the 2 is $\frac{2}{100}$.

PRACTICE

Fill in the blanks below. *Remember:* **The decimal places are *to the right* of the decimal point.**

1 In the number 4.05, what place is the 5 in? _____

2 What is the value of the 1 in 5.129? _____

3 What is the value of the 3 in 0.035? _____

4 The number 0.06 is six _____.

5 The number 0.9 is nine _____.

6 The number 0.1, written in words,

is _____.

7 In digits, eight hundredths is _____.

In a decimal name, the place value you say is the place value of the digit farthest to the right. For example, 0.083 is "eighty-three thousandths."

8 0.85 is eighty-five _____.

9 0 .015 is fifteen _____.

When you say a number that has a whole number as well as digits in the decimal places, the word **and** is used for the decimal point. For example, the number 5.64 is "five *and* sixty-four hundredths."

Write each number below in digits.

10 six and three tenths _____

11 twelve and two tenths _____

Comparing Decimal Numbers

To compare two decimal numbers, start by lining up the decimal points. Then, moving from left to right, compare the digits.

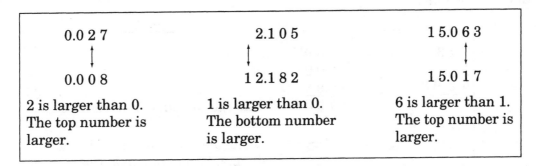

0.0 2 7	2.1 0 5	1 5.0 6 3
↕	↕	↕
0.0 0 8	1 2.1 8 2	1 5.0 1 7
2 is larger than 0. The top number is larger.	1 is larger than 0. The bottom number is larger.	6 is larger than 1. The top number is larger.

As another example, compare 0.1 and 0.008. You can think of 0.1 as 0.100. In the tenths place, 1 is larger than 0. That means 0.1 is larger than 0.008. With decimals, any nonzero digit *close to* a decimal point represents a larger value than any digit *further from* the decimal point.

PRACTICE

Circle the smaller number in each problem.

1 0.003   0.87

2 0.13   0.098

3 0.00899   0.051

4 1.75   1.032

5 0.76   0.13

6 0.02   0.005

Circle the smallest number in each problem.

7 0.51   0.09   1.3

8 11   0.19   1.023

9 1.6   1.8   1.05

10 0.13   0.07   0.002

Arrange each set of numbers from smallest to largest.

11 0.1, 1, 0.01, 11

_____ , _____ , _____ , _____

12 0.032, 0.023, 0.75, 0.15

_____ , _____ , _____ , _____

13 8.7, 0.75, 2.3

_____ , _____ , _____

Suppose that a decimal number begins with zero and a decimal point, and then has four digits. Using the digits below, rearrange them to make the *smallest* decimal number possible.

14 0 9 5 1   _____

15 7 0 1 3   _____

Adding Decimals

To add the numbers **4.3, 2,** and **0.93,** start by writing the numbers in column form with the decimal points lined up. For the whole number **2,** write a decimal point just to the right of the ones place.

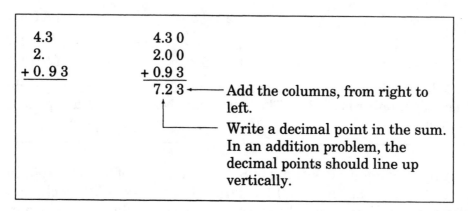

Write a zero in each empty column. Then add the columns, from right to left. When you are finished adding each column, write a decimal point in your answer. All the decimal points should line up vertically.

PRACTICE

Add and then check your answers. Be sure to start with the problem written in column form.

1
| 5.6 | 0.14 | $1.00 | 1.09 | 2. | 1. |
| + 3. | + 0.32 | + 1.60 | + 0.05 | + 0.62 | + 0.03 |

2   0.19 + 3 = _____   0.12 + 0.07 = _____   5 + 0.264 = _____

3   0.075 + 0.123 = _____   6 + 0.012 = _____   0.52 + 1.3 = _____

4   0.238 + 6 = _____   52.8 + 0.15 = _____   23.02 + 0.095 = _____

Subtracting Decimals

To subtract decimals, start by writing the numbers in column form, with the decimal points lined up. Write a zero in each empty column, then subtract. The decimal point in the difference should line up with the other decimal points.

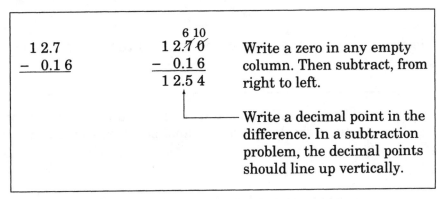

PRACTICE

Subtract and then check your answers. Start with each problem in column form. If the difference in a column is zero, be sure to write that zero.

1
7.9	0.74	$9.00	0.10	2.	7.56
− 5.	− 0.71	− 1.75	− 0.05	− 0.62	− 0.08

2   $0.25 - 0.15 =$ ____ $0.8 - 0.06 =$ ____ $0.09 - 0.03 =$ ____

3   $6.6 - 0.13 =$ ____ $1.09 - 0.25 =$ ____ $11.3 - 0.113 =$ ____

4   $0.2 - 0.078 =$ ____ $0.903 - 0.16 =$ ____ $6 - 2.556 =$ ____

Multiplying Decimals

When you multiply decimals, at first you ignore the decimals points. Write the problem with the numbers lined up at the right. After you multiply, count the number of digits to the right of each decimal point. That is the **number of decimal places** for each number.

The number of decimal places in the product must equal the *sum* of the decimal places in the numbers you multiplied.

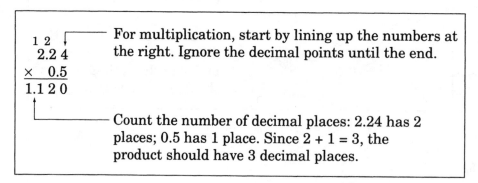

For multiplication, start by lining up the numbers at the right. Ignore the decimal points until the end.

Count the number of decimal places: 2.24 has 2 places; 0.5 has 1 place. Since 2 + 1 = 3, the product should have 3 decimal places.

PRACTICE

Multiply. Then check your answers.

1
$$\begin{array}{r} 0.3 \\ \times\, 0.6 \\ \hline \end{array}$$
$$\begin{array}{r} 1.3 \\ \times\, 0.2 \\ \hline \end{array}$$
$$\begin{array}{r} \$5.40 \\ \times\quad 5 \\ \hline \end{array}$$
$$\begin{array}{r} \$0.75 \\ \times\quad 3 \\ \hline \end{array}$$

2
$$\begin{array}{r} 2.5 \\ \times\, 0.3 \\ \hline \end{array}$$
$$\begin{array}{r} 0.305 \\ \times\quad 5 \\ \hline \end{array}$$
$$\begin{array}{r} \$3.33 \\ \times\quad 15 \\ \hline \end{array}$$
$$\begin{array}{r} \$0.15 \\ \times\quad 3 \\ \hline \end{array}$$

3    $4.4 \times 3 =$ _____ $0.66 \times .2 =$ _____ $9.6 \times 6.2 =$ _____

4   Omar has 12 boards. Each is 2.3 feet long. How many feet of board does he have in all?

6   What number is three hundredths times 42?

5   What number is two tenths times 32?

7   There are 5.2 million people living in a large city. Three tenths of them are children. How many children live in that city?

_____ million

In the problem below, 0.15 has two decimal places and 0.3 has one decimal place. That means the product must have 2 + 1 = 3 decimal places. To write the product, you have to write a zero *to the left* of the nonzero digits.

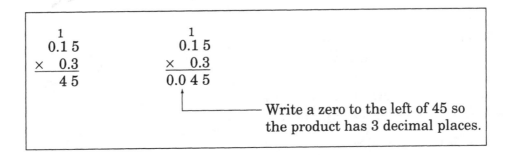

Write a zero to the left of 45 so the product has 3 decimal places.

Multiply, then check your answers.

8
| 0.12 | 0.21 | 4.2 | 0.75 | 0.06 | 0.31 |
| × 0.3 | × 0.4 | × 0.5 | × 0.4 | × 0.7 | × 0.2 |

9
| 0.12 | 0.25 | 0.22 | 0.34 | 0.05 | 0.07 |
| × 0.15 | × 0.14 | × 0.25 | × 0.2 | × 0.09 | × 0.2 |

10 What is fifteen hundredths times 0.35?

11 One box of cereal weighs 0.75 pound. How much do 50 boxes of cereal weigh?

12 In 1992, there were about 0.5 times as many divorces as weddings in the U.S. There were 1.2 million weddings. About how many divorces were there?

13 Erica has 3 feet of gold braid. She must give one-half of it to her sister. How much gold braid will Erica give her sister?

Multiplying Decimals

61

Solving Word Problems

Below, circle the letter for the number sentence that shows how to solve each problem.

1   Mia is ordering two jackets through the mail. One weighs 1.3 kilograms (kg). The other weighs 3.15 kilograms. How much will her order weigh altogether?

     A   3.15 kg – 1.3 kg = ☐ kg

     B   3.15 kg × 1.3 = ☐ kg

     C   3.15 kg + 1.3 kg = ☐ kg

     D   Not enough information is given.

2   Ground round is 4 dollars a pound. You buy 4.6 pounds. How much will you pay? (Ignore the tax.)

     F   4 dollars × 4.6 = ☐ dollars

     G   4 dollars – 4.6 lb = ☐ dollars

     H   4 dollars + 4.6 lb = ☐ dollars

     J   Not enough information is given.

3   Experts predict that girls born in 1970 will live an average of 74.7 years. They predict that, on average, boys born that year will live 61.7 years. On average, how much longer will the women live than the men?

     A   74.7 + 61.7 = ☐

     B   61.7 – 74.7 = ☐

     C   74.7 – 61.7 = ☐

     D   Not enough information is given.

Solve each word problem below. Use estimation to check your answers.

4   An average of 3.13 inches of rain falls in June. An average of 3.05 inches falls in July. Which month gets more rain?

     _____

5   Which is longer, 4.5 feet of copper pipe or 4.28 feet of copper pipe?

     _____

6   Coco drove to New York. When she began the trip, her mileage odometer was at 11,545.6 miles. When she got to New York, it read 11,966.9 miles. How many miles did she drive?

     _____

7   There are 3.485 million people living in Los Angeles, and there are 1.007 million people living in Dallas. How many more people live in Los Angeles than in Dallas?

     _____ million

8   There are 14.6 million people living in and around New York City. Twice that many live in and around Tokyo. The population of the Tokyo area is how many people?

     _____ million

Decimals Skills Practice

Circle the letter for the correct answer to each problem. Try crossing out unreasonable answers before you start to work on each problem.

1 $15 + 3.1 =$ _____

- **A** 153.1
- **B** 15.31
- **C** 1.81
- **D** 4.6
- **E** None of these

2 $5.1 - 0.06 =$ _____

- **F** 5.4
- **G** 4.4
- **H** 5.04
- **J** 4.04
- **K** None of these

3 $0.04 \times 4 =$ _____

- **A** 16
- **B** 1.6
- **C** 0.16
- **D** 4.04
- **E** None of these

4 $\$8.22 \times 3 =$ _____

- **F** $24.22
- **G** $24.55
- **H** $24.66
- **J** $11.66
- **K** None of these

5 $2.3 + 6.4 =$ _____

- **A** 8.43
- **B** 8.34
- **C** 8.7
- **D** 23.64
- **E** None of these

6
$$45.2$$
$$- 3.9$$

- **F** 0.413
- **G** 49.1
- **H** 42.3
- **J** 41.3
- **K** None of these

7 Julio earns an average of $31.50 a night in tips. At this rate, about how much will he earn in 5 nights?

- **A** $50.00
- **B** $75.00
- **C** $150.00
- **D** $175.00

8 Jerry earns $14.80 per hour. Yesterday he worked for 5.25 hours. Which number sentence should he use to estimate how much he earned yesterday?

- **F** $14.00 \times 5 = \square$
- **G** $15.00 \times 5 = \square$
- **H** $14.00 \times 6 = \square$
- **J** $15.00 \times 6 = \square$

9 Link's cleaning bill came to $18.46. He paid with a $20.00 bill. How much change should he receive?

- **A** $1.44
- **B** $2.44
- **C** $1.54
- **D** $2.54

10 Obami is buying shades for her windows. They cost $6.00 each, and she has 7 windows. How much will she spend?

- **F** $76.00
- **G** $49.00
- **H** $42.00
- **J** $13.00

11

$0.3 + 0.7 = \underline{\hspace{1cm}}$

A 1
B 0.1
C 1.1
D 0.9
E None of these

12

$\begin{array}{r} 0.3 \\ \times\ 6 \\ \hline \end{array}$

F 0.18
G 18
H 1.8
J 1.08
K None of these

13

$\begin{array}{r} 1 \\ -\ 0.2 \\ \hline \end{array}$

A 8
B 0.8
C 0.08
D 80
E None of these

14

$\begin{array}{r} 98.0 \\ -\ 5.15 \\ \hline \end{array}$

F 46.5
G 93.95
H 92.85
J 93.85
K None of these

15

$1.402 \times 0.2 = \underline{\hspace{1cm}}$

A 0.284
B 28.04
C 0.2804
D 2.804
E None of these

16

$\begin{array}{r} 5.1 \\ \times\ 2.3 \\ \hline \end{array}$

F 11.73
G 2.55
H 25.5
J 117.3
K None of these

17

$25 + 3.5 = \underline{\hspace{1cm}}$

A 60
B 6
C 0.6
D 28.5
E None of these

18

$100 \times 6.178 = \underline{\hspace{1cm}}$

F 617.8
G 0.06178
H 6.17800
J 6,178
K None of these

19

$35.0 - 0.1706 = \underline{\hspace{1cm}}$

A 34.8204
B 34.9304
C 34.8294
D 34.824
E None of these

20 Which set of decimal numbers is in order from least to greatest?

F 1, 1.07, 0.8, 0.09
G 0.8, 1, 0.09, 1.07
H 0.8, 0.09, 1, 1.07
K 0.09, 0.8, 1, 1.07

21 A number begins with the digit 0 and a decimal point. Then it uses the digits 1, 6, 5, and 0, each exactly once. What is the smallest decimal that can be written?

A 0.0651
B 0.6510
C 0.1560
D 0.0156

22 What is 19.079 rounded to the nearest tenth?

F 20
G 19
H 19.1
J 19.08

Fractions

Numerators and Denominators

A number such as $\frac{1}{2}$ (one-half) is called a **fraction.** A fraction represents a part of something. If you divide a cake into 6 equal pieces, each piece is $\frac{1}{6}$ of the cake.

In a fraction, the number on the bottom tells how many parts the object or amount has been divided into. The bottom number is called the **denominator.**

 In this figure, $\frac{1}{8}$ is shaded.

Tell the fraction of each shape that is shaded.

1   2   3 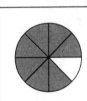   4

_____   _____   _____   _____

The top number in a fraction is called the **numerator.** It refers to the number of parts that are shown or shaded.

 In this figure, $\frac{7}{8}$ is shaded.

Tell the fraction of each shape that is shaded.

5   6   7   8

_____   _____   _____   _____

Below each fraction is a circle. Divide each circle into the correct number of parts, and shade that fraction of the circle.

9 $\frac{1}{2}$ 10 $\frac{5}{8}$ 11 $\frac{1}{6}$ 12 $\frac{2}{2}$

Fractions and Division

Write a fraction to represent each amount described below. Remember: The top number refers to the number of parts shown or shaded.

1 There are 100 cents in a dollar. What part of a dollar is 27 cents?

_____

2 There are 1,760 yards in a mile. What part of a mile is 153 yards?

_____

3 There are 52 weeks in a year. Eliza had three weeks of vacation this year. What part of the year did she take off?

_____

4 Wendy has 23 living relatives, and 15 of them are on her husband's side of the family. What fraction of Wendy's relatives are on her husband's side?

_____

5 There are 9 workers in the office. This year, 2 of them will retire. What fraction of the workers will retire?

_____

6 Terri has saved $31.00 to buy a new dress. The dress costs $175.00. What fraction of the cost has she saved so far?

_____

7 Jorge mows lawns for 15 dollars apiece. He saves 1 dollar for every lawn he mows. What fraction of his income does Jorge save?

_____

8 Darnell earns $500 a week. He pays $156 a week in taxes. What fraction of his pay is taken out for taxes?

_____

The fraction $\frac{1}{3}$ can be used when a large group is divided into 3 smaller groups. You can use division to find the size of each smaller group.

$\frac{1}{3}$ of 312: $312 \div 3 = 104$

$\frac{1}{6}$ of 600: $600 \div 6 = 100$

Find each fraction described below.

9 Find $\frac{1}{4}$ of 24.

10 Find $\frac{1}{7}$ of 14.

11 Find $\frac{1}{8}$ of 32.

12 Find $\frac{1}{3}$ of 15.

13 Find $\frac{1}{2}$ of 100.

14 Find $\frac{1}{9}$ of 90.

15 Find $\frac{1}{5}$ of 25.

16 Find $\frac{1}{10}$ of 50.

17 Find $\frac{1}{6}$ of 36.

Reducing a Fraction to Simplest Terms

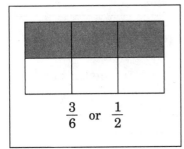

$$\frac{3}{6} \quad \text{or} \quad \frac{1}{2}$$

There are several ways to name the fraction that represents this figure. If you think of the figure as being divided into six parts, then $\frac{3}{6}$ is shaded. If you think of it as being divided into two parts, then $\frac{1}{2}$ is shaded. The fractions $\frac{3}{6}$ and $\frac{1}{2}$ have the same value. They are called **equivalent fractions.**

When you reduce a fraction, you change it into an equivalent fraction with smaller numbers. It is easier to understand and use fractions when they have been reduced. **To reduce a fraction, divide both the top and the bottom by the same number.**

The fraction $\frac{3}{9}$ can be reduced further by dividing the numerator 3 and the denominator 9 by 3:

$$\frac{3}{9} = \frac{3 \div 3}{9 \div 3} = \frac{1}{3}$$

Here is a problem: **Reduce $\frac{12}{30}$.**

Both the numerator 12 and the denominator 30 can be evenly divided by 2, so

$$\frac{12}{30} = \frac{12 \div 2}{30 \div 2} = \frac{6}{15}$$

Now, the new numerator 6 and the new denominator 15 can be evenly divided by 3, so

$$\frac{6}{15} = \frac{6 \div 3}{15 \div 3} = \frac{2}{5}$$

No number can evenly divide both 2 and 5, so $\frac{2}{5}$ cannot be reduced any more. It is in **lowest terms** or **simplest terms.** This is also called **reducing a fraction.**

PRACTICE

Reduce each fraction to lowest terms.

> *Hint:* Any fraction with a 1 in the numerator is in lowest terms.

1 $\frac{8}{16} = $ _____

2 $\frac{10}{50} = $ _____

3 $\frac{15}{75} = $ _____

4 $\frac{20}{100} = $ _____

5 $\frac{33}{99} = $ _____

6 $\frac{40}{100} = $ _____

7 $\frac{12}{24} = $ _____

8 $\frac{40}{80} = $ _____

9 $\frac{5}{25} = $ _____

10 $\frac{4}{16} = $ _____

11 $\frac{10}{30} = $ _____

12 $\frac{9}{36} = $ _____

Finding Factors

An important step in reducing a fraction is to find a number that evenly divides other numbers. In general, a number that divides a larger number is called a **factor** of the larger number.

Problem: What are the factors of 12?

Think: $12 \div 1 = 12$, so 1 and 12 are factors of 12.
         $12 \div 2 = 6$, so 6 and 2 are factors of 12.
         $12 \div 3 = 4$, so 4 and 3 are factors of 12.

PRACTICE

List all the factors of each number. There is one blank space for every possible factor.

1   The factors of 10 are 1 , _____, _____, and 10.

2   The factors of 9 are 1, _____, and 9.

Hint: Any number that ends in 0, 2, 4, 6, or 8 can be divided by 2.

3   The factors of 14 are 1, _____, _____, and 14.

4   The factors of 15 are 1, _____, _____, and 15.

5   The factors of 16 are 1 , _____, _____, _____, and 16.

6   The factors of 20 are 1, _____, _____, _____, _____, and 20.

Hint: any number that ends in 0 can be divided by 10, 5, and 2.

7   Write all three factors of 25. _____

Write the factors of each number.

8   Write all six factors of 50. _____

9   Write all six factors of 18. _____

10   Write all nine factors of 100. _____

11   Write all eight factors of 24. _____

The numbers 30 and 36 have several factors that are the same: 1, 2, 3, and 6. The largest of those factors is 6, so 6 is called the **greatest common factor** of 30 and 36.

Use the lists above to find the greatest common factor for each pair of numbers.

12	12 and 24	_____	15	9 and 15	_____	18	100 and 24	_____
13	15 and 20	_____	16	16 and 20	_____	19	14 and 18	_____
14	25 and 50	_____	17	18 and 9	_____	20	10 and 50	_____

Fractions Equal To 1 and Fractions Greater Than 1

So far, you have worked with fractions such as $\frac{4}{5}$ or $\frac{1}{3}$, where the top number is smaller than the bottom number. These fractions are called **proper fractions.** If the top number in a fraction is equal to or larger than the bottom number, such as $\frac{5}{4}$, $\frac{7}{6}$, and $\frac{9}{9}$, the fraction is called an **improper fraction.**

Here are some other numbers that are greater than one: $3\frac{3}{4}$, $6\frac{1}{2}$, $1\frac{1}{4}$, and so on. Each of these is a **mixed number.** A mixed number is the sum of a whole number and a proper fraction.

PRACTICE

Tell whether each fraction is greater than 1, equal to 1, or less than 1. *Hint:* Fractions such as $\frac{3}{3}$, $\frac{5}{5}$, and $\frac{7}{7}$ are equal to 1.

1	$\frac{7}{8}$ _____	**3**	$\frac{3}{5}$ _____	**5**	$\frac{9}{5}$ _____	**7**	$\frac{3}{10}$ _____
2	$\frac{8}{8}$ _____	**4**	$2\frac{1}{5}$ _____	**6**	$4\frac{1}{4}$ _____	**8**	$\frac{25}{82}$ _____

To change an improper fraction to a whole number, start by dividing the top number by the bottom number. If there is no remainder, your answer is a whole number.	If there is a remainder, then your answer is a mixed number. For the fraction, the numerator is the remainder and the denominator is the denominator of the original improper fraction.
$$\frac{8}{2} = 8 \div 2 = 4$$	$$\frac{7}{5} = 7 \div 5 = 1\text{ r }2 \text{ } or \text{ } \frac{7}{5} = 1\frac{2}{5}$$

Write each improper fraction as a whole number or as a mixed number.

9	$\frac{16}{8} =$ ____	**12**	$\frac{12}{5} =$ ____	**15**	$\frac{17}{5} =$ ____	**18**	$\frac{19}{9} =$ ____
10	$\frac{21}{7} =$ ____	**13**	$\frac{8}{5} =$ ____	**16**	$\frac{19}{3} =$ ____	**19**	$\frac{26}{5} =$ ____
11	$\frac{9}{8} =$ ____	**14**	$\frac{15}{3} =$ ____	**17**	$\frac{17}{10} =$ ____	**20**	$\frac{35}{8} =$ ____

Comparing Fractions

Here are some of the things you have already learned about fractions.

♦ If the numerator is larger than the denominator, the fraction is greater than one.
♦ If the numerator equals the denominator, then the fraction equals one.
♦ If the numerator is less than the denominator, then the fraction is less than one.

Here are two more ideas about fractions.

If two fractions have the same denominator, such as $\frac{3}{4}$ and $\frac{1}{4}$, then the fraction with the larger numerator has the greater value.

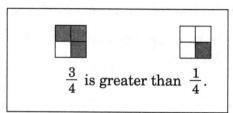

$\frac{3}{4}$ is greater than $\frac{1}{4}$.

If two fractions have the same numerator, such as $\frac{1}{3}$ and $\frac{1}{6}$ or such as $\frac{2}{5}$ and $\frac{2}{7}$, then the fraction with the *smaller* denominator has the greater value.

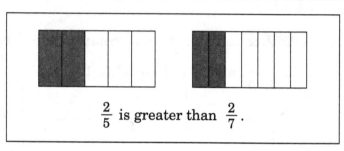

$\frac{2}{5}$ is greater than $\frac{2}{7}$.

PRACTICE

Circle the smaller number in each problem.

1	$\frac{5}{4}$	$\frac{3}{5}$		5	$\frac{1}{4}$	$\frac{1}{5}$		9	$\frac{4}{5}$	$\frac{7}{7}$
2	$\frac{3}{3}$	$\frac{2}{3}$		6	$\frac{1}{3}$	$\frac{1}{7}$		10	$\frac{1}{4}$	$\frac{1}{6}$
3	$\frac{1}{3}$	$\frac{1}{7}$		7	$\frac{2}{5}$	$\frac{4}{5}$		11	$1\frac{1}{5}$	$\frac{3}{5}$
4	$\frac{3}{5}$	$\frac{2}{5}$		8	$\frac{3}{5}$	$\frac{3}{7}$		12	$3\frac{1}{4}$	$5\frac{1}{4}$

Arrange these fractions from smallest to largest.

13   $\frac{4}{5}$ $\frac{7}{5}$ $\frac{2}{5}$ $\frac{5}{5}$ _____   16 $1\frac{1}{2}$ $1\frac{1}{3}$ $1\frac{1}{5}$ _____

14   $\frac{1}{2}$ $\frac{1}{4}$ $\frac{1}{5}$ $\frac{1}{3}$ _____   17 $\frac{3}{4}$ $\frac{3}{5}$ $\frac{3}{7}$ _____

15   $\frac{2}{9}$ $\frac{7}{9}$ $\frac{5}{9}$ $\frac{6}{9}$ _____

Comparing Fractions

Adding Fractions

To add two fractions that have the same denominator, use that same denominator and add the two numerators.

$$\frac{2}{5} + \frac{1}{5} = \frac{2+1}{5} \longleftarrow \text{Add the numerators.}$$
$$\phantom{\frac{2}{5} + \frac{1}{5}} = \frac{3}{5} \phantom{\frac{2+1}{5}} \longleftarrow \text{Use the same denominator.}$$

Reduce the sum if it is not in lowest terms.

$$\frac{1}{4} + \frac{1}{4} = \frac{1+1}{4} = \frac{2}{4} \longleftarrow \text{Add the numerators.}$$
$$\phantom{\frac{1}{4} + \frac{1}{4} = \frac{1+1}{4}} \longleftarrow \text{Use the same denominator.}$$
$$\frac{2}{4} = \frac{2 \div 2}{4 \div 2} = \frac{1}{2} \longleftarrow \text{Reduce the fraction to lowest terms.}$$

PRACTICE

Add the fractions below. You do not have to reduce the sums.

1 $\quad \frac{1}{6} + \frac{4}{6} = $ _____

2 $\quad \frac{2}{7} + \frac{1}{7} = $ _____

3 $\quad \frac{4}{9} + \frac{1}{9} = $ _____

4 $\quad \frac{1}{5} + \frac{1}{5} = $ _____

5 $\quad \frac{3}{8} + \frac{2}{8} = $ _____

6 $\quad \begin{array}{r} \frac{2}{9} \\ + \frac{2}{9} \\ \hline \end{array}$

Add the fractions below. Be sure each sum is reduced to lowest terms.

7 $\quad \frac{1}{6} + \frac{1}{6} = $ _____

8 $\quad \frac{2}{6} + \frac{2}{6} = $ _____

9 $\quad \frac{1}{10} + \frac{1}{10} = $ _____

10 $\quad \frac{1}{6} + \frac{2}{6} = $ _____

11 $\quad \frac{1}{8} + \frac{1}{8} = $ _____

12 $\quad \begin{array}{r} \frac{2}{15} \\ + \frac{3}{15} \\ \hline \end{array}$

First, write each sum as an improper fraction. Then write the sum as a whole number or as a mixed number.

13 $\quad \frac{3}{5} + \frac{2}{5} = $ _____

14 $\quad \frac{3}{8} + \frac{5}{8} = $ _____

15 $\quad \frac{5}{7} + \frac{9}{7} = $ _____

16 $\quad \frac{7}{9} + \frac{7}{9} = $ _____

17 $\quad \frac{4}{5} + \frac{3}{5} = $ _____

18 $\quad \begin{array}{r} \frac{3}{6} \\ + \frac{4}{6} \\ \hline \end{array}$

Subtracting Fractions

In the subtraction problem below, the fractions have the same denominator. To subtract the fractions, use that same denominator and subtract the numerators.

$$\frac{7}{8} - \frac{3}{8} = \frac{7-3}{8} \quad \longleftarrow \text{Subtract the numerators.}$$
$$\longleftarrow \text{Use the same denominator.}$$
$$= \frac{4}{8}$$
$$\frac{4}{8} = \frac{4 \div 4}{8 \div 4} = \frac{1}{2} \quad \longleftarrow \text{Reduce the fraction.}$$

For a problem such as $\frac{6}{7} - \frac{6}{7} = \underline{\ ?\ }$, the difference is $\frac{6}{7} - \frac{6}{7} = \frac{6-6}{7} = \frac{0}{7}$.

Since zero over any (nonzero) number is zero, the result is $\frac{6}{7} - \frac{6}{7} = \frac{0}{7} = 0$

PRACTICE

Subtract the fractions below. Be sure each difference is reduced to lowest terms.

1 $\frac{7}{12} - \frac{5}{12} =$ _____

2 $\frac{8}{10} - \frac{6}{10} =$ _____

3 $\frac{5}{7} - \frac{5}{7} =$ _____

4 $\frac{3}{5} - \frac{2}{5} =$ _____

5 $\frac{10}{12} - \frac{8}{12} =$ _____

6 $\begin{array}{r} \frac{3}{6} \\ - \frac{1}{6} \\ \hline \end{array}$

7 $\frac{7}{15} - \frac{2}{15} =$ _____

8 $\frac{19}{20} - \frac{4}{20} =$ _____

9 $\frac{25}{30} - \frac{10}{30} =$ _____

10 $\frac{3}{4} - \frac{1}{4} =$ _____

11 $\frac{5}{10} - \frac{3}{10} =$ _____

12 $\begin{array}{r} \frac{7}{8} \\ - \frac{3}{8} \\ \hline \end{array}$

13 $\frac{15}{15} - \frac{5}{15} =$ _____

14 $\frac{18}{20} - \frac{3}{20} =$ _____

15 $\frac{45}{50} - \frac{20}{50} =$ _____

16 $\frac{15}{30} - \frac{10}{30} =$ _____

17 $\frac{9}{24} - \frac{3}{24} =$ _____

18 $\begin{array}{r} \frac{9}{10} \\ - \frac{1}{10} \\ \hline \end{array}$

Multiplying Fractions

To multiply two fractions, multiply the numerators together. Then multiply the denominators together.

$$\frac{2}{4} \times \frac{3}{6} = \frac{2 \times 3}{4 \times 6} = \frac{6}{24}$$ ⟵ Multiply the numerators and multiply the denominators.

$$\frac{6}{24} = \frac{6 \div 6}{24 \div 6} = \frac{1}{4}$$ ⟵ Reduce the product.

PRACTICE

Multiply the fractions below. Be sure each product is reduced to lowest terms.

1 $\frac{2}{3} \times \frac{1}{4} =$ _____

2 $\frac{4}{5} \times \frac{1}{2} =$ _____

3 $\frac{1}{4} \times \frac{4}{5} =$ _____

4 $\frac{1}{6} \times \frac{1}{4} =$ _____

5 $\frac{3}{9} \times \frac{1}{3} =$ _____

6 $\frac{3}{6}$
$\times \frac{2}{6}$

7 $\frac{1}{9}$
$\times \frac{5}{5}$

8 $\frac{5}{7}$
$\times \frac{1}{3}$

9 $\frac{1}{7}$
$\times \frac{5}{5}$

10 $\frac{1}{3} \times \frac{3}{2} \times \frac{1}{2} =$ _____

11 $\frac{1}{2} \times \frac{1}{2} \times \frac{1}{3} =$ _____

12 $\frac{1}{7} \times \frac{3}{7} =$ _____

13 $\frac{1}{5} \times \frac{2}{5} =$ _____

14 $\frac{2}{4} \times \frac{1}{2} \times \frac{1}{2} =$ _____

15 $\frac{5}{9}$
$\times \frac{1}{9}$

16 $\frac{1}{2}$
$\times \frac{2}{6}$

17 $\frac{1}{4}$
$\times \frac{3}{4}$

18 $\frac{1}{3}$
$\times \frac{2}{5}$

$\frac{1}{3}$ of $\frac{1}{2}$
is the same as
$\frac{1}{3} \times \frac{1}{2}$

19 What is $\frac{1}{3}$ of $\frac{1}{4}$?

20 What is $\frac{1}{4}$ of $\frac{1}{5}$?

21 What is $\frac{1}{2}$ of $\frac{1}{4}$?

22 What is $\frac{1}{5}$ of $\frac{1}{2}$?

Canceling Before You Multiply

To multiply $\frac{2}{5}$ times $\frac{3}{4}$, notice that the factors in the numerator and denominator have a common factor 2:

$$\frac{2}{5} \times \frac{3}{4} = \frac{2 \times 3}{5 \times 4} = \frac{?}{\underline{}}$$

A shortcut is to reduce the fractions before you multiply. This shortcut is called **canceling.** To cancel, look for a common factor in the numerator and the denominator. Divide both numbers by the common factor. Then multiply as usual.

$$\frac{2}{5} \times \frac{3}{4} = \frac{\overset{1}{\cancel{2}} \times 3}{5 \times \underset{2}{\cancel{4}}}$$

—— Divide 2 into the numerator and divide 2 into the denominator.

$$= \frac{3}{10}$$

—— Then multiply the numerators and multiply the denominators.

In this problem, you can cancel two times.

$$\frac{3}{4} \times \frac{2}{3} \times \frac{1}{2} = \frac{3 \times \overset{1}{\cancel{2}} \times 1}{4 \times 3 \times \underset{1}{\cancel{2}}}$$

Cancel by the common factor 2.

$$= \frac{\overset{1}{\cancel{3}} \times \overset{1}{\cancel{2}} \times 1}{4 \times \underset{1}{\cancel{3}} \times \underset{1}{\cancel{2}}}$$

Cancel by the common factor 3.

$$= \frac{1}{4}$$

PRACTICE

Simplify the problems below by canceling. Then finish the multiplication.

1. $\dfrac{2}{\cancel{8}_{1}} \times \dfrac{\cancel{3}^{1}}{5} = \dfrac{2}{5}$

2. $\dfrac{1}{\cancel{4}_{2}} \times \dfrac{\cancel{2}^{1}}{3} = \dfrac{b}{6}$

3. $\dfrac{\cancel{2}^{1}}{5} \times \dfrac{3}{\cancel{4}_{2}} = \dfrac{3}{10}$

4. $\dfrac{\cancel{3}^{1}}{8} \times \dfrac{2}{\cancel{3}_{1}} \times \dfrac{1}{\cancel{5}_{1}} = \underline{}$

5. $\dfrac{4}{5} \times \dfrac{1}{2} \times \dfrac{1}{3} = \underline{}$

6. $\dfrac{\cancel{5}^{1}}{6} \times \dfrac{1}{\cancel{10}_{2}} = \dfrac{1}{12}$

7. $\dfrac{\cancel{2}^{1}}{3} \times \dfrac{1}{\cancel{2}_{1}} = \dfrac{1}{3}$

8. $\dfrac{\cancel{4}^{2}}{9} \times \dfrac{1}{\cancel{2}_{1}} = \dfrac{2}{9}$

9. $\dfrac{\cancel{4}^{1}}{5} \times \dfrac{3}{\cancel{8}_{2}} = \dfrac{3}{10}$

10. $\dfrac{1}{3} \times \dfrac{3}{6} \times \dfrac{2}{4} = \underline{}$

Multiplying a Fraction by a Whole Number

To multiply a whole number and a fraction, start by writing the whole number as an improper fraction. To do that, simply write the whole number over one. Then you can multiply the fractions.

$$2 \times \frac{3}{4} = \frac{2}{1} \times \frac{3}{4} = \frac{2 \times 3}{1 \times 4} = \frac{6}{4}$$

Reduce: $\frac{6}{4} = \frac{6 \div 2}{4 \div 2} = \frac{3}{2} = 1\frac{1}{2}$

Canceling first:

$$2 \times \frac{3}{4} = \frac{2}{1} \times \frac{3}{4} = \frac{\overset{1}{\cancel{2}} \times 3}{1 \times \underset{2}{\cancel{4}}} = \frac{3}{2} = 1\frac{1}{2}$$

PRACTICE

Find each product below. Reduce all answers to lowest terms. You will use fewer steps if you cancel before you multiply.

1   $8 \times \dfrac{2}{4} =$ _____

2   $3 \times \dfrac{2}{3} =$ _____

3   $2 \times \dfrac{1}{4} =$ _____

4   $5 \times \dfrac{3}{5} =$ _____

5   $3 \times \dfrac{2}{9} =$ _____

6   $7 \times \dfrac{1}{5} =$ _____

7   $2 \times \dfrac{1}{3} =$ _____

8   $3 \times \dfrac{3}{8} =$ _____

9   $4 \times \dfrac{3}{4} =$ _____

10   $\dfrac{3}{8} \times 4 =$ _____

11   $\dfrac{9}{10} \times 2 =$ _____

12   $\dfrac{3}{21} \times 7 =$ _____

13   $\dfrac{1}{3} \times 15 =$ _____

14   $\dfrac{4}{5} \times 3 =$ _____

Rounding a Fraction or a Mixed Number

When a problem includes fractions or mixed numbers, you can estimate by rounding to whole numbers.

A fraction less than $\frac{1}{2}$ rounds to zero. A proper fraction $\frac{1}{2}$ or more rounds to 1.

Fraction	Rounds to
$\frac{1}{5}$	0
$\frac{6}{7}$	1
$\frac{1}{3}$	0
$\frac{3}{4}$	1

To round a mixed number to the nearest whole number, round the fraction to 1 or to 0. Then add that value to the whole-number part of the mixed number.

Mixed Number	Rounds to
$2\frac{1}{3}$	2
$3\frac{3}{4}$	4
$5\frac{1}{2}$	6

PRACTICE

Round each fraction or mixed number to the nearest whole number.

1   $1\frac{4}{5}$ _2_

2   $3\frac{1}{4}$ _3_

3   $5\frac{1}{5}$ _5_

4   $6\frac{1}{3}$ _6_

5   $2\frac{1}{2}$ _2_

6   $\frac{3}{4}$ _1_

7   $4\frac{5}{6}$ _5_

8   $9\frac{1}{10}$ _9_

Below, estimate the answer to each problem by rounding the fractions.

9   $1\frac{3}{4} + 1\frac{1}{2} =$ _2_

10   $2\frac{1}{3} - 1\frac{1}{4} =$ _4_

11   $3\frac{2}{5} + 5\frac{3}{8} =$ _9_

12   $4\frac{1}{2} \times 2\frac{1}{2} =$ _7_

13   $10\frac{1}{6} - 4\frac{6}{7} =$ _15_

14   $2\frac{1}{2} + 2\frac{1}{4} =$ _5_

Solving Word Problems

Solve each word problem below. Reduce all your answers to simplest terms.

1 You are building a chair. The plans call for $\frac{1}{2}$-inch nails. Your nails are $\frac{3}{4}$-inch long. Are your nails too long or too short?

too loNS

2 There are 16 ounces in a pound. What fraction of a pound is 8 ounces?

$\frac{1}{2}$

3 Mari needs 2 yards of scrap cloth. She finds five $\frac{1}{2}$-yard pieces. Does she have enough cloth?

Yes

4 Your recipe calls for 300 grams of chocolate. You have 200 grams. What fraction of the chocolate do you have?

$\frac{2}{3}$

5 Of the 125 people in the community choir, 22 of them actually come from other communities. What fraction of the choir comes from other communities?

6 John is doubling a cake recipe. The recipe calls for $\frac{1}{4}$ cup of butter. How much butter should John use?

2/4

7 Martina ordered carpet that is $\frac{7}{8}$ of an inch thick. She ordered padding that is $\frac{5}{8}$ of an inch thick. When they are put together, how thick will they be?

ABout 1½ IN

8 Yesterday it rained $\frac{3}{4}$ of an inch. Today it rained $\frac{1}{4}$ of an inch. How much rain has there been altogether?

1 IN

9 Ruby is growing tomato plants in her garden. Last week one plant was $2\frac{3}{4}$ feet tall. This week it is $3\frac{1}{3}$ feet tall. How much has it grown?

1 $\frac{2}{3}$

10 Melinda is going to buy a crystal figure from a catalog. Her catalog says that the crystal cat is $\frac{3}{8}$ of an inch tall. It says that the crystal baby is $\frac{3}{4}$ of an inch tall. Which figure is taller?

The BABy

11 Your cookie recipe calls for $\frac{2}{4}$ cup of white sugar and $\frac{1}{4}$ cup of honey. How much sweetener do you need in all?

3/4

Comparing Numbers Using Ratios

Each of the following five statements is a *ratio:*
>   There are 4 men for every woman in the group.
>   There is 1 murder committed every 5 minutes.
>   He scores one free throw out of every three he attempts.
>   She drove 60 miles per hour.
>   The chocolates cost $8.00 per pound.

A **ratio** is a comparison of two numbers. Ratios can be written in three ways:

1 to 2   **1 : 2** $\dfrac{1}{2}$

The third way, using a fraction, is the most useful.

The five ratios from the top of this page are shown at the right, each in fraction form. Notice that the word *per* refers to "1 unit."

$\dfrac{4 \text{ men}}{1 \text{ woman}}$	$\dfrac{1 \text{ murder}}{5 \text{ minutes}}$	$\dfrac{1 \text{ score}}{3 \text{ attempts}}$	$\dfrac{60 \text{ miles}}{1 \text{ hour}}$	$\dfrac{\$8.00}{1 \text{ pound}}$

PRACTICE

Write each ratio as a fraction. Use a number and a unit in the numerator and the denominator.

1   Use 1 cup of ginger ale for every 3 cups of fruit juice. $\frac{1}{3}$

2   There are 2 teachers in the school for every 31 students. $\frac{2}{31}$

3   Every 100 miles you drive, take a 15-minute rest. $\frac{15}{100}$

4   There are 4 quarts in a gallon. $\frac{1}{4}$

5   On the highway, Janet's car gets 22 miles per gallon. $\frac{22}{1}$

6   Three cans of soup cost one dollar. $\frac{1}{3}$

7   It costs $12.00 per family. $\frac{1}{2}$

You can reduce a ratio the same way you reduce a fraction. Write each ratio in lowest terms.

8   There are 10 managers in the company and 40 regular employees. What is the ratio of managers to regular employees? $\frac{1}{4}$

9   The Cougars lost 8 games and won 4. What was the ratio of games lost to games won? $\frac{1}{2}$

10   Patrice works 15 hours a week babysitting and 25 hours a week driving a bus. What is the ratio of hours she spends on these two jobs? $\frac{1}{5}$

11   There are 16 cans of soda in Coco's refrigerator and 4 cans of juice. What is the ratio of soda cans to juice cans? $\frac{4}{16}$

Finding a Unit Rate

In a ratio, if the number in the denominator is 1, that ratio is called a **unit rate.**
Examples of unit rates are miles per gallon, meters per second, and dollars per pound.

To calculate a unit rate, such as miles per hour:

1. Put the information you have in ratio form. The word that follows *per* goes into the denominator.
2. Simplify the ratio until the bottom number is 1.

Problem: At a constant speed, it took Glenn 6 hours to drive 360 miles. How fast did he drive?

Write a ratio: $\dfrac{360 \text{ miles}}{6 \text{ hours}}$

Simplify it:

$$\dfrac{360 \text{ miles}}{6 \text{ hours}} = \dfrac{360 \div 6}{6 \div 6} = \dfrac{60 \text{ miles}}{1 \text{ hour}}$$

Answer: 60 miles per hour

PRACTICE

Find the unit rate for each problem.

1 Raul drove 480 miles. He used 24 gallons of gas. How many miles did he drive per gallon of gasoline? *27.1*

2 Eldridge paid $30.00 for 6 ounces of perfume. How much did the perfume cost per ounce? *$5*

3 You can buy a box of 10 candy bars for $8.00. What is the price per candy bar? *1≥5*

4 Felicia is a singer who performs 10 times a week. Her weekly pay is $750.00. How much does she earn per performance? *750÷10=$75*

5 Pablo rode 36 miles on his bike. It took two hours. About how many miles did he ride per hour? *18*

6 Randy works 40 hours a week. She is paid $480 per week. How much does she make an hour? *10.80*

7 It took Helen 3 days to drive 1,200 miles. About how many miles did she drive each day? *1200÷3=400*

8 You can buy 3 pounds of frozen corn for $5.40. What is the cost per pound? *5.40÷3=1.60*

Finding a Unit Rate

Writing Proportions

If you write an equation with one ratio equal to another ratio, such as $\dfrac{3}{6} = \dfrac{6}{12}$, you have a **proportion.** When you think of ratios as fractions, then the two ratios in a proportion are **equivalent fractions.**

Proportions: $\quad \dfrac{3 \text{ dollars}}{6 \text{ cans}} = \dfrac{6 \text{ dollars}}{12 \text{ cans}} \qquad \dfrac{10 \text{ meters}}{3 \text{ seconds}} = \dfrac{30 \text{ meters}}{9 \text{ seconds}}$

In this problem you know three of the numbers in a proportion, and you need to find the fourth number.

Problem

The cost of 12 roses is 10 dollars. How much would 24 roses cost?

Write a ratio of the two units, dollars and roses: $\dfrac{\text{dollars}}{\text{roses}}$.

Then write ratios using the numbers in the problem, with "?" as the unknown number. Be sure all the numerators use the numbers for dollars and all the denominators use the numbers for roses.

$$\frac{\text{dollars}}{\text{roses}} = \frac{10}{12} = \frac{?}{24}$$

You can get 24 in the denominator by multiplying 10 and 12 by 2:

$$\frac{10}{12} = \frac{10 \times 2}{12 \times 2} = \frac{20}{24}$$

Then look at the proportion $\dfrac{?}{24} = \dfrac{20}{24}$. The cost for 24 roses is 20 dollars.

PRACTICE

Below, write each proportion in fraction form. *Remember:* **Use like labels in the numerators and denominators.**

1 Every 2 hours the factory produces 25 cars. In 8 hours, it produces 100 cars. _____

2 Yvonne types 60 words a minute. In 3 minutes, she types 180 words. _____

3 12 eggs cost $1.00. At that rate, 6 eggs cost $0.50. _____

Write each of these problems as a proportion. Use the symbol "?" for the missing number in each problem.

4 On a map, each inch stands for 50 miles. The towns of Sedalia and Crocker are 2 inches apart on the map. How many miles apart are the two towns? _____

5 It takes Dana 2 days to make 50 pots. How many pots can she make in 20 days? _____

For problems 6–8, tell how to solve each proportion.

6 $\dfrac{16 \text{ ounces}}{\$2.00} = \dfrac{? \text{ ounces}}{\$4.00}$ 32

To solve, multiply 16 and 2 by _____ .

7 $\dfrac{60 \text{ seconds}}{1 \text{ minute}} = \dfrac{? \text{ seconds}}{30 \text{ minutes}}$ ×60

To solve, multiply 60 and 1 by _____ .

8 $\dfrac{3 \text{ teaspoons}}{1 \text{ tablespoon}} = \dfrac{? \text{ teaspoons}}{6 \text{ tablespoons}}$

To solve, multiply 3 teaspoons by _6_ .

Solve each proportion below.

9 $\dfrac{3 \text{ women}}{5 \text{ men}} = \dfrac{\boxed{9} \text{ women}}{15 \text{ men}}$

10 $\dfrac{20 \text{ blocks}}{2 \text{ kilometers}} = \dfrac{\boxed{20} \text{ blocks}}{10 \text{ kilometers}}$

11 $\dfrac{8 \text{ ounces}}{1 \text{ dollar}} = \dfrac{\boxed{24} \text{ ounces}}{3 \text{ dollars}}$

12 $\dfrac{50 \text{ dollars}}{3 \text{ months}} = \dfrac{100 \text{ dollars}}{\boxed{6} \text{ months}}$

13 $\dfrac{44 \text{ miles}}{3 \text{ gallons}} = \dfrac{88 \text{ miles}}{\boxed{6} \text{ gallons}}$

Write a proportion for each problem below. Then find the missing number in your proportion.

14 Biscuits are marked 3 cans for $1.95. How much would 12 cans cost?

15 On average, Ronny hits the dunking booth target 3 times out of 10. If he buys 50 balls, about how many times will he hit the target?

16 On average, the appliance store sells five television sets every two hours. How many television sets do they sell in one 8-hour shift?

17 Grant needs 2 cans of fertilizer for every 50 square yards of lawn. His lawn is about 400 square yards. How much fertilizer does he need?

18 Ms. Leyland can seat 4 students at each table in her room. She has 8 tables. How many students can she seat?

19 A bag of candy contains eighty pieces, and it costs $3.50. How much would it cost to buy two hundred forty pieces of candy?

Writing Proportions

Fractions Skills Practice

Circle the letter for the correct answer to each problem. Reduce all fractions to lowest terms.

1

$$\frac{5}{7} + \frac{3}{7} = \underline{\qquad}$$

A $\frac{2}{7}$

B $\frac{4}{7}$

C 8

D $1\frac{1}{7}$

E None of these

2

$$\frac{5}{12} - \frac{1}{12} = \underline{\qquad}$$

F $\frac{1}{4}$

G $\frac{1}{6}$

H $\frac{1}{12}$

J $\frac{1}{3}$

K None of these

3

$$7 \times \frac{4}{7} = \underline{\qquad}$$

A 7

B 4

C $1\frac{3}{4}$

D $1\frac{4}{7}$

E None of these

4

$$\frac{3}{4} \times \frac{2}{5} = \underline{\qquad}$$

F $\frac{5}{9}$

G $\frac{2}{3}$

H $\frac{1}{4}$

J $\frac{2}{5}$

K None of these

Study this recipe. Then do Numbers 5 through 7.

Root Beer Syrup

2 tablespoons root beer extract

2 cups sugar

$\frac{1}{4}$ teaspoon brewing yeast

4 cups warm water

Dissolve the yeast in a small bit of warm water. After 10 minutes, mix it with the remaining ingredients. Pour the liquid into clean soda bottles. Seal the bottles tightly, and leave them in a warm spot for 3–5 days. Store them for 2 weeks more in a cool spot before drinking. Makes enough syrup for 2 liters of root beer.

5 You have a 6-tablespoon packet of root beer extract. What fraction of the packet should you use?

A $\frac{1}{2}$ B $\frac{1}{3}$ C $\frac{1}{4}$ D $\frac{1}{6}$

6 What amount of brewing yeast do you need?

F between $\frac{1}{2}$ and $\frac{3}{4}$ teaspoon

G between 0 and $\frac{1}{5}$ teaspoon

H between $\frac{1}{3}$ and $\frac{1}{2}$ teaspoon

J between 0 and $\frac{1}{2}$ teaspoon

7 If you need 4 cups of water to make enough syrup for 2 liters of root beer, how much water do you need to make enough syrup for 8 liters of root beer?

A 4 cups C 0 cups
B 8 cups D 16 cups

8

$$\frac{25}{5} = \underline{\qquad}$$

   A 5   **C** 10
   B $\frac{1}{5}$ **D** 25
   E None of these

9

$$\begin{array}{r} \frac{3}{5} \\ + \frac{4}{5} \\ \hline \end{array}$$

   F $\frac{7}{10}$ **H** $1\frac{2}{5}$
   G $2\frac{2}{5}$ **J** $\frac{12}{25}$
   K None of these

10

$$\frac{6}{7} - \frac{2}{7} = \underline{\qquad}$$

   A $\frac{4}{14}$ **C** $\frac{3}{7}$
   B 4   **D** $\frac{4}{7}$
   E None of these

11

$$\frac{1}{2} \times \frac{1}{5} \times \frac{1}{8} = \underline{\qquad}$$

   F $\frac{1}{15}$ **H** $\frac{3}{15}$
   G $\frac{1}{80}$ **J** $\frac{3}{80}$
   K None of these

12

$$\frac{1}{6} \times \frac{7}{5} = \underline{\qquad}$$

   A $\frac{8}{30}$ **C** $\frac{7}{11}$
   B $\frac{7}{30}$ **D** $\frac{8}{11}$
   E None of these

13

$$\begin{array}{r} \frac{4}{6} \\ - \frac{1}{6} \\ \hline \end{array}$$

   F $\frac{1}{2}$ **H** $\frac{1}{4}$
   G $\frac{1}{3}$ **J** 3
   K None of these

14

$$\frac{7}{10} - \frac{7}{10} = \underline{\qquad}$$

   A 1   **C** 7
   B 0   **D** $\frac{1}{10}$
   E None of these

15

$$\frac{12}{12} = \underline{\qquad}$$

   F 12 **H** 0
   G 2   **J** 1
   K None of these

16 Which of these number sentences is true?

   A $\frac{1}{2}$ is less than $\frac{1}{3}$

   B $\frac{6}{5}$ is less than 1

   C $\frac{11}{12}$ is less than 1

   D $\frac{4}{5}$ is less than $\frac{1}{2}$

17 Which set of fractions is in order from least to greatest?

   F $\frac{1}{12}, \frac{1}{2}, \frac{1}{3}$

   G $\frac{1}{2}, \frac{1}{12}, \frac{1}{3}$

   H $\frac{1}{2}, \frac{1}{3}, \frac{1}{12}$

   J $\frac{1}{12}, \frac{1}{3}, \frac{1}{2}$

18 What fraction of a dollar is 15 cents?

   A $\frac{1}{15}$

   B $\frac{15}{10}$

   C $\frac{15}{100}$

   D $\frac{100}{15}$

19 There are 24 hours in a day. What fraction of a day is 8 hours?

   F $\frac{1}{2}$

   G $\frac{1}{4}$

   H $\frac{1}{3}$

   J $\frac{1}{5}$

Fractions Skills Practice

Data Interpretation

Reading a Table

Tables and graphs are useful ways to organize information and show many numbers. In order to understand a table or a graph, always begin by reading the title and the headings. They explain the relationships shown in a table or graph.

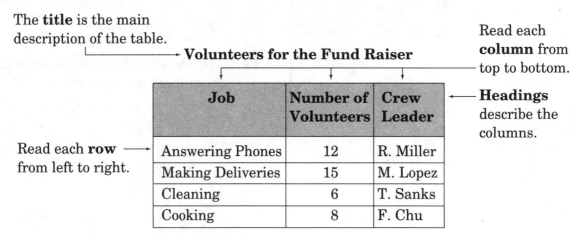

The **title** is the main description of the table.

Read each **row** from left to right.

Read each **column** from top to bottom.

Headings describe the columns.

Volunteers for the Fund Raiser

Job	Number of Volunteers	Crew Leader
Answering Phones	12	R. Miller
Making Deliveries	15	M. Lopez
Cleaning	6	T. Sanks
Cooking	8	F. Chu

PRACTICE

Use the table above to answer each question.

1 What does this table describe?

  A telephones
  B different types of jobs
  C cooking
  D people who are helping out at a fund raiser

2

Making Deliveries	15	M. Lopez

What is this part of the table called?

  F a column
  G a row
  H a heading
  J the title

3 What does this column in the table tell you?

12
15
6
8

  A how many hours each person worked
  B how many phones there are
  C how many people have offered to help with each job
  D There is no way to tell.

4 What is described in the third column of the table?

  F who is leading each group of workers
  G cleaning
  H cooking
  J the names of the people who wrote the table

For this table, suppose you want to know the cost of renting a regular video for 5 days. Find the row for "regular videos," and find the column for "5 days." Then find the box, or cell, where they meet. That box has the information you want.

Look down the column for **5 days**.

Look along the row for **regular videos**. →

Rental Fees

	Number of Days		
	1	3	5
Regular Videos	$1.00	$2.00	$3.00
New Releases	$2.50	$5.00	$7.50
Video Games	$1.75	$3.25	$6.50
Computer Software	$1.75	$3.25	$6.50

PRACTICE

Use the table above to answer these questions.

5 How much do you pay to rent a regular video for 3 days? _____

6 How much does it cost to rent a video game for 5 days? _____

7 You rent a video game for 3 days and a regular video for 3 days. How much do you pay? _____

8 What can you rent for $1.35 or less? _____

9 What amount is the highest rental fee this store charges? _____

10 What is the fee for renting a new release for 3 days? _____

11 *Revenge of the Zombies* just came out on video, and this store has it. Which row tells the cost to rent this movie?

_____

12 Your children found a computer game they want to rent. Which row tells the rental fee for the game?

_____

13 Circle the letter for each piece of information that is in the table.

A how much the store will charge if you lose a video
B whether this store is more expensive than the place where you usually rent videos
C how much it would cost to rent a video
D whether this store sells snacks
E whether this store has foreign films

Using Numbers in a Table

You can use the numbers in a table to make comparisons. Start by finding each number. Then:

♦ You can find the **difference** between two numbers by subtracting.
♦ You can see how **many times larger** one number is than another by dividing.
♦ You can find **what fraction** one number is of another by forming a ratio.

Years with the Company

Employee	Years of Service
Anne Rodriguez	25
Clinton Reed	10
Linda Hansen	2
David Blume	10
Heidi Hunt	5
Myumi Kino	8

Question
How many times longer has Anne Rodriguez been with the company than Heidi Hunt?
Solution
Ann Rodriguez: 25 years; Heidi Hunt: 5 years
$$25 \div 5 = 5$$

Question
Fill in the blank: Linda Hansen has been with the company _?_ times as long as David Blume.
Solution
Linda Hansen: 2 years; David Blume: 10 years
$$\frac{2}{10} = \frac{1}{5}$$

PRACTICE

Use the table above to fill in the blanks.

1   Who has been with this company the shortest time? _____

2   Who has been with the company longer, Myumi Kino or David Blume? _____

3   How much longer has Clinton Reed been with the company than Heidi Hunt? _____

4   Clinton Reed has been with the company _?_ times as long as Linda Hansen. _____

5   Myumi Kino has been with the company _?_ times as long as Linda Hansen. _____

Use the table below to answer questions 6–8.

Average School Scores on a State Test

School	Reading	Writing	Math
Lincoln	70	64	35
Thorpe	73	69	42
Tubman	86	73	57
Wiley	72	63	32
Roosevelt	64	55	32

6   Which school had the highest math score? _____

7   Which school had the lowest writing score? _____

8   Lincoln students' reading scores were _?_ times higher than their math scores. _____

Using a Price List

A menu or price list is a common type of table. To use a menu or price list, find what you want to buy. Then look for the price for that item. Do not get confused if you are looking for several prices. Write the price for each item. Then do any figuring.

Copying Charges

	price per page
black-and-white copies	$0.08
double-sided copies	$0.05
hand-fed copies	$0.25
color copies	$1.49
laminating	$2.50
spiral binding	add $1.50 per packet
plastic covers	$2.50 each

Reminder:
- To put different amounts together, add the numbers.
- To compare one amount to another, subtract or divide the numbers.
- To take something away from something else, subtract the numbers.
- To solve problems involving groups of objects, multiply or divide the numbers

PRACTICE

Use the price list above to answer these questions.

1 How much does it cost for 20 black-and-white copies? _____

2 How much more do color copies cost per page than black-and-white copies? _____

3 You make one color copy and have it laminated. How much do you spend? _____

4 You have 20 double-sided copies made. Then you put the copies together using one spiral binding. How much do you spend? _____

5 How much does it cost to make ten color copies and buy one plastic cover? _____

6 Laura has 10 copies to make. How much will the job cost if they must be fed into the copy machine by hand? _____

7 Jason has only 80 cents to spend. How many black-and-white copies can he make? _____

8 How much can you save per page if you print your black-and-white copies on two sides? _____

Median and Mode

Ahmad received the test scores shown at the right.

Math:	72, 82, 72, 82, 85, 72, 91
Science:	87, 81, 79, 85, 89, 92

There are several ways to describe a *typical* math score or science score. One way is to list the scores in order and find the middle value. That is called the **median** of the scores.

Math Scores

72   72   72 $\boxed{82}$ 82   85   91

There are 7 scores, an odd number. For an odd number of values, the median is the middle value, which is

Science Scores

79   81 $\boxed{85\quad 87}$ 89   92

$\boxed{86}$

There are 6 scores, an even number. The median is halfway between 85 and 87, which is 86.

PRACTICE

Find the median for each set of numbers.

1   Gwen did five speed tests in typing class. Her times, in minutes, were 2, 3, 2, 5, and 4. What was her median time?

_____

2   Ron sold 15 raffle tickets, Jan sold 35, Claudio sold 23, Nunzio sold 47, and Erin sold 23. What was the median number of tickets sold?

_____

3   On a state licensing test, Jean and her friends got the following scores: 65, 78, 63, 78, 97, 42, and 53. What was their median score?

_____

Another way to describe a typical score is to identify the number that appears most often. This core is called the **mode.** For Ahmad's math scores, the mode is 72. For Ahmad's science scores, no value appears more than one time so there is no mode.

4   Look again at Gwen's typing times: 2, 3, 2, 5, and 4 minutes. What is the mode for this set of numbers?

_____

5   In question 2, what is the mode for the number of tickets sold?

_____

6   In question 3, what is the mode of the set of test scores?

_____

Finding an Average

Another way to describe a *typical* score is an **average** or **mean**.

To find an average of a set of number:
First, *add* all the numbers in the set.
Then, *divide* that sum by the total number of values.

Years with the Company

Employee	Years of Service
Anne Rodriguez	25
Clinton Reed	10
Linda Hansen	2
David Blume	10
Heidi Hunt	5
Myumi Kino	8

Question

What is the average number of years of service for these six people?

Solution

Add all the numbers: 25 + 10 + 2 + 10 + 5 + 8 = 60
Divide by 6, the number of people: 60 ÷ 6 = 10

The average number of years of service is 10 years.

PRACTICE

1 There are three women in Jenny's carpool. Their ages are 52 years, 29 years, and 39 years. What is the average age of the women in the carpool?

_____

2 Here are the number of minutes of coffee breaks Nathan took this week: 20, 30, 10, 10, 30.

What was the average amount of time Nathan took each day?

_____

3 The first five times Riley drove to his new job it took 37 minutes, 45 minutes, 39 minutes, 41 minutes, and 43 minutes. What is the average time it took Riley to drive to his new job?

_____

4 The Bargains Galore Catalog Store keeps a record of how long its four telephone operators take to complete a call. Complete the table.

Bargains Galore Telephone Orders

Operator	Minutes Spent on Each Call						Median	Mode	Mean
	1	2	3	4	5	6			
A	2	2	4	6	3	1	2.5		3
B	3	3	5	4	5	X		None	
C	2	2	1	2	3	2			
D	6	5	5	6	8	X			

Graphs

A graph presents information. You can find many graphs in newspapers, magazines, and reports. A graph can show a lot of information quickly and clearly.

Here are three different types of graphs:

A **circle graph** divides a circle into slices or wedges to show parts of a whole.

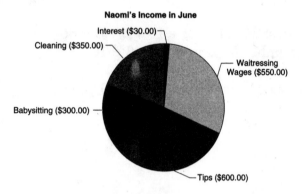

Naomi's Income in June

1 How many categories of income are shown?

A **line graph** uses points or dots to show values. Lines connecting the points show rising or falling values.

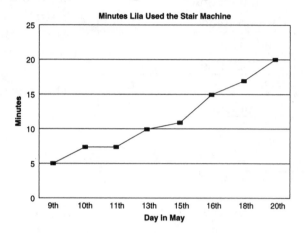

Minutes Lila Used the Stair Machine

2 How many days of information are shown?

A **bar graph** uses thick bars to represent numbers.

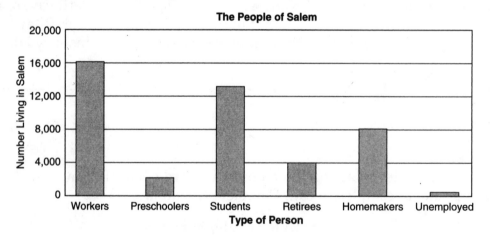

The People of Salem

3 How many categories of people are shown?

Reading a Circle Graph

A circle graph shows how a whole is divided into parts. It is divided into sections, like a pie, and each section stands for a fraction of the total. The larger the section, the larger the fraction.

In a circle graph it is easy to compare each section to the whole. In the circle graph at the right, the section for Naomi's income from tips represents about one-third of her total income in June.

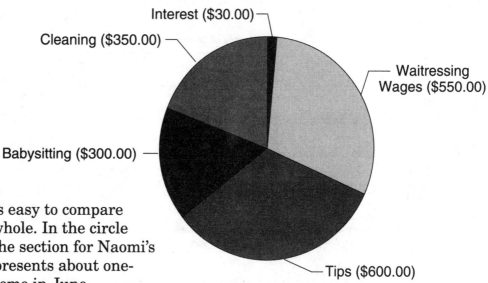

Naomi's Income in June

Interest ($30.00)

Cleaning ($350.00)

Waitressing Wages ($550.00)

Babysitting ($300.00)

Tips ($600.00)

PRACTICE

Use the graph above to answer these questions.

1   The money Naomi got for babysitting represents about what fraction of her June income?

    A $\frac{1}{2}$ **C** $\frac{2}{3}$

    B $\frac{1}{6}$ **D** $\frac{3}{4}$

2   Which category represents the biggest portion of Naomi's income?

    F waitressing wages   **H** cleaning
    G babysitting   **J** tips

3   Together, Naomi's wages and tips made up about what fraction of her June income?

    A $\frac{1}{10}$ **C** $\frac{3}{10}$

    B $\frac{1}{5}$ **D** $\frac{2}{3}$

4   Did Naomi make more money from tips or from cleaning?

    F tips   **G** cleaning fees

5   If Naomi becomes a manager at the restaurant, her wages would double but she would not get any tips. Would she make more money or less money if she is a manager instead of a waitress?

    A more money   **B** less money

Reading a Bar Graph

A bar graph uses thick lines, or bars, to represent values. The longer the bar, the larger the number.

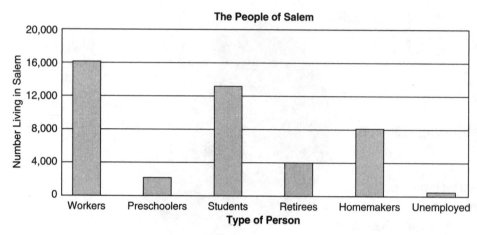

Question
How many people living in Salem are homemakers?

Solution
Find the bar for homemakers.
The top of that bar represents 8,000.
There are 8,000 homemakers in Salem.

If the top of a bar does not line up exactly with a number, you can estimate the value of the bar. For example, the end of the bar for students lies about one-fourth of the way between 12,000 and 16,000. Since 13,000 is one-fourth of the way between 12,000 and 16,000, then a good estimate is 13,000 students.

PRACTICE

1 How many people living in Salem are retirees? _____

2 Which of these is the best estimate of how many people living in Salem are unemployed?

    F 4,000   **H** 1,000
    G 0   **J** 3,000

3 Which of these is the best estimate of how many preschoolers live in Salem?

    A 4,000   **C** 1,000
    B 3,000   **D** 2,000

4 Are there more workers or students living in Salem? _____

5 About how many homemakers are there for every retiree in Salem? (*Think:* There are _?_ times as many homemakers as retirees.) _____

6 There are 44,000 people living in Salem. What fraction of the population has a job outside the home? _____

7 Which of the following people would find this graph most useful?

    A a politician trying to figure out what the people of Salem want and need

    B a worker trying to decide when to retire

    C someone trying to find out whether the schools in Salem are good

Reading a Line Graph

A line graph uses points or dots to show values. The numbers along one side of the graph show the values of the points on the graph.

On the graph, lines between the points show whether the values are rising or falling. So line graphs show trends and changes in amounts.

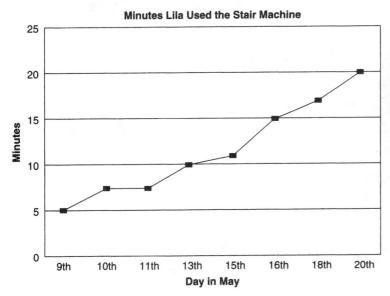

Question
On May 11, how many minutes did Lila spend on the stair machine?

Solution
Find May 11 along the bottom of the graph. Then find the point that represents May 11.
The point for May 11 is halfway between 5 and 10 minutes. The value that is halfway between 5 and 10 is $7\frac{1}{2}$, so Lila spent about $7\frac{1}{2}$ minutes on the stair machine on May 11.

PRACTICE

Use the graph above to answer these questions.

1 On May 9, how long was Lila on the stair machine? _____

2 On May 13, how long was Lila on the stair machine? _____

3 On May 18, about how many minutes did Lila spend on the stair machine? _____

4 How much more time was Lila on the stair machine on May 20 than on May 16? _____

5 By May 20th, Lila was spending __?__ times as long on the machine as she spent on May 9th. _____

6 On which two dates did Lila spend the same amount of time on the stair machine?

_____

7 Which of these is the best estimate of the *average* time Lila spent on the stair machine?

A 5 minutes   **C** 8 minutes
B 12 minutes   **D** 17 minutes

8 Which trend is shown in this graph?

F Lila is tired of the machine.
G Lila is spending more and more time on the stair machine.
H If Lila is sad, she spends no time on the stair machine.

Using a Graph

The graph below tells you fractions of a total. To find the number for each section, you can multiply each fraction by the total.

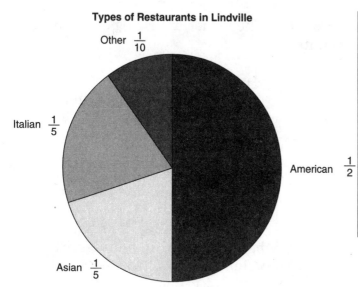

Types of Restaurants in Lindville

Other $\frac{1}{10}$

Italian $\frac{1}{5}$

Asian $\frac{1}{5}$

American $\frac{1}{2}$

Question
Suppose there are 50 restaurants in Lindville. How many of them serve Asian food?

Solution
Find $\frac{1}{5}$ of 50.

$$\frac{1}{5} \times 50 = \frac{1}{5} \times \frac{50}{1} = \frac{50}{5} = \frac{10}{1} = 10$$

PRACTICE

Use the circle graph above to answer these questions. *Remember:* **There are 50 restaurants altogether.**

1   How many American restaurants are there in Lindville? _____

2   How many of Lindville's restaurants are not Italian, Asian, or American? _____

3   How many of the restaurants in Lindville are either Italian or Asian? (*Hint:* Add the fractions before you multiply.) _____

4   Suppose that in 1997, about $\frac{3}{5}$ of the restaurants in Lindville served fast food. How many fast food restaurants did Lindville have in 1997? _____

Data Interpretation Skills Practice

Hakim sells appliances. This graph shows how much he sold last week. Study the graph. Then do Numbers 1 through 6.

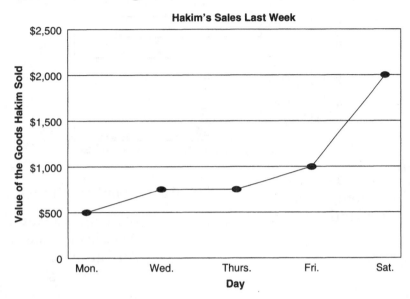

Hakim's Sales Last Week

1   On what two days, if any, did Hakim sell the same value of goods?

    A   Tuesday and Wednesday
    B   Wednesday and Thursday
    C   Thursday and Friday
    D   No two days were the same.

2   Between what two days did Hakim's sales increase the most?

    F   Monday and Wednesday
    G   Wednesday and Thursday
    H   Thursday and Friday
    J   Friday and Saturday

3   Which of these is the best estimate of Hakim's average sales per day last week?

    A $500
    B $600
    C $1,000
    D $1,750

4   Which of these is the best estimate of the value of the goods Hakim sold on Thursday?

    F $510
    G $750
    H $850
    J $900

5   Hakim sold twice as much this Saturday as he did last Saturday. How much did he sell this Saturday?

    A $1,000
    B $2,000
    C $3,000
    D $4,000

6   What was the difference in the value of Hakim's sales on Monday and Saturday?

    F $500
    G $1,000
    H $1,500
    J $2,500

This graph shows types of families for American children. Study the graph. Then do Numbers 7 through 11.

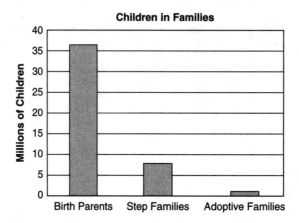

Children in Families

7 Which of these is the best estimate of how many U.S. children live in adoptive families?

A 8 million
B 1 million
C 3 million
D 1 out of 10

8 Which of these is the best estimate of the total number of children living in step or adoptive families?

F 9 million
G 6 million
H 11 million
J 5 million

9 There are about 46 million children in the United States. What fraction of them live with their birth parents?

A about $\frac{1}{4}$

B about $\frac{1}{3}$

C about $\frac{1}{2}$

D about $\frac{3}{4}$

10 About $\frac{1}{7}$ of the American children living in step families are African-American. About how many African-American children live in step families?

F 1 million
G 7,000
H   70
J 5,000

11 Which of the following statements is supported by this graph?

A About $\frac{1}{2}$ of U.S. children live in step families.
B There are about four times as many U.S. children living with birth parents as there are living in step families.
C The number of step families in the U.S. is rising.
D There are about twice as many step families in the U.S. as adoptive families.

Algebra

Patterns

A key to understanding math is to see and use patterns. When you describe a pattern using words and symbols, you are using the math ideas called **algebra.**

PRACTICE

1 Which of the following describes the pattern of squares in the box below?

A 1 white, 1 gray, 1 white, 1 black   **C** 1 white, 2 gray, 1 white, 2 black
B 1 white, 2 gray, 2 white, 2 black   **D** 1 white, 2 gray, 1 black, 2 gray

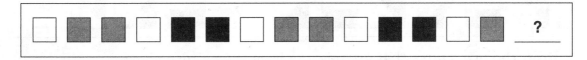

2 Describe the next square in the pattern above.

_____

3 Which of the following describes the pattern in the box below?

A point up, point down   **C** point up, point right, point down, point right
B point up, point right   **D** point up, point right, point down, point left

4 Describe the missing pattern in the box above.

_____

5

The missing figures in the pattern above should be divided into _?_ and _?_ parts.

_____

Answer each problem.

6 Describe the pattern in the box below.

_____

7 Circle the letter for the missing figure in the pattern above.

A   B   C

8

Draw the two figures that are missing from the pattern above.

9

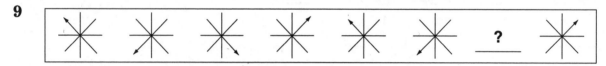

Draw the figure that is missing from the pattern above.

10

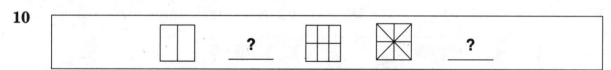

Draw the two figures that are missing from the pattern above.

11

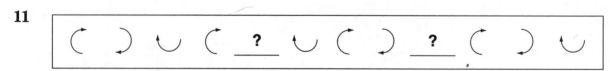

Draw the two figures that are missing from the pattern above.

Finding Number Patterns

You can create a pattern with numbers as well as with shapes. Here are some number pattern problems.

PRACTICE

For each problem, fill in the blanks.

1   Count down by threes,
    starting with sixteen:
    16, 13, 10   _____ _____ _____

2   Add 6 to each number,
    starting with three:
    3, 9, 15   _____ _____ _____

3   Add 5 to each number,
    starting with
    three: 3, 8, 13   _____ _____ _____

4   Multiply each number
    by 2, starting with
    two: 2, 4, 8   _____ _____ _____

5   Starting with 1, multiply
    each number by 2 and
    then add 1: 1, 3, 7   _____ _____ _____

6   1, 5, 9, 13, 17, 21
    This pattern shows
    counting by _?_ .   _____

7   1, 12, 23, 34, 45, 56
    In this pattern, each
    new number is the last
    number plus _?_ .   _____

8   25, 22, 19, 16, 13, 10
    In this pattern, each
    new number is the last
    number minus _?_ .   _____

9   1, 3, 9, 27
    In this pattern, each
    new number is the
    last number times _?_ .   _____

10   0, 7, 14, 21, 28
     In this pattern, each new
     number is the last _?_ .   _____

11   15, 30, 45, 60, 75, 90
     In this pattern, each new
     number is the last _?_ .   _____

12   29, 24, 19, 14, 9, 4
     In this pattern, each new
     number is the last _?_ .   _____

13   30, 24, 18, 12, 6, 0
     In this pattern, each new
     number is the last _?_ .   _____

14   Fill in the missing number in
     this pattern:
     3, 6, 9, 12, 15, _?_ , 21, 24, 27   _____

15   Fill in the missing numbers:
     50, 45, 40, _?_ , 30,
     25, _?_ , 15   ____ ____

16   Fill in the missing numbers:
     5, 7, 9, _?_ , _?_ , 15, 17, 19   ____ ____

17   The next pattern is a little
     different. Try to fill in the
     missing number:
     1, 3, 3, 2, 4, 4, 5, 6, 6, 7, 8, _?_   _____

18   Fill in the missing number:
     1, 2, 4, 5, 7, 8, 10,
     11, 13, 14, _?_   _____

19   Fill in the missing number:
     1, 3, 2, 4, 3, 5, 4, 6, 5, _?_   _____

Patterns in Number Sentences

The number sentences below are incomplete. Use what you know about number facts to answer each question.

PRACTICE

Write +, −, ×, or ÷ for each number sentence.

Sample 1
The sentence $2 \,\square\, 3 = 5$ needs the symbol + because $2 \,\boxed{+}\, 3 = 5$.

1 $\quad 4 \,\square\, 6 = 10$

2 $\quad 5 \,\square\, 3 = 2$

3 $\quad 9 \,\square\, 3 = 27$

4 $\quad 1 \,\square\, 9 = 10$

5 $\quad 11 \,\square\, 7 = 18$

6 $\quad 27 \,\square\, 5 = 22$

7 $\quad 35 \,\square\, 5 = 7$

8 $\quad 12 \,\square\, 6 = 2$

9 $\quad 13 \,\square\, 13 = 26$

10 $\quad 156 \,\square\, 16 = 140$

Write the number that makes each number sentence true.

Sample 2
For $5 + \square = 9$, write 4 because $5 + \boxed{4} = 9$.

11 $\quad 3 + \square = 7$

12 $\quad 5 + \square = 13$

13 $\quad 9 + \square = 19$

14 $\quad 33 + \square = 45$

15 $\quad 12 - \square = 8$

Hint: To solve $12 - \square = 8$, write $12 - 8 = \square$.

16 $\quad 15 - \square = 6$

17 $\quad 52 - \square = 40$

18 $\quad 75 - \square = 53$

Hint: To solve $7 \times \square = 14$, write $14 \div 7 = \square$.

19 $\quad 7 \times \square = 14$

20 $\quad 6 \times \square = 60$

21 $\quad 5 \times \square = 60$

Patterns in Number Sentences

Some Basic Number Properties

Here are the number properties you have learned in this book. The number properties are the basic rules of algebra. Review the rules in each box, and use them for the problems below.

The order of the numbers in a problem *does not* matter in addition and multiplication problems but *does* matter in subtraction and division problems.

$5 + 7$ and $7 + 5$ have the *same* value.
5×7 and 7×5 have the *same* value.
$7 - 5$ and $5 - 7$ have *different* values.
$15 \div 5$ and $5 \div 15$ have *different* values.

Addition and subtraction are **inverse operations.** Each one will undo the other. Similarly, multiplication and division are inverse operations, so each one will undo the other.

$+$ $2 + 3 = 5$, so $5 - 3 = 2$.
$-$ $7 - 3 = 4$, so $4 + 3 = 7$.
$\times$ $6 \times 7 = 42$, so $42 \div 7 = 6$.
$\div$ $50 \div 10 = 5$, so $5 \times 10 = 50$.

There are two **properties of one.** First, if you multiply or divide a number by one, you do not change the value.
     $15 \times 1 = 15$ $20 \div 1 = 20$

Second, if you divide any (nonzero) number by itself, the quotient is one.
$15 \div 15 = 1$

There are several **properties of zero.** If you add or subtract zero from a number, you do not change the value.
     $15 + 0 = 15$ $20 - 0 = 20$

If you subtract a number from itself, the difference is zero. $15 - 15 = 0$

If you multiply any number by zero, the product is zero. $20 \times 0 = 0$

If you divide zero by any (nonzero) number, the quotient is zero. $0 \div 15 = 0$

An expression like $15 \div 0$ has no meaning in math.

PRACTICE

Write +, −, ×, or ÷ in the box for each number sentence.

1 $123 \ \boxed{} \ 123 = 1$

2 $14 \ \boxed{} \ 14 = 0$

3 $0 \ \boxed{} \ 32 = 0$

Below, write the sign +, −, ×, or ÷ and a number.

4 $769 + 84 = 853$

    $853 \ \boxed{} = 769$

5 $488 + 28 = 516$

    $516 \ \boxed{} = 28$

6 $12 \times 11 = 132$

    $132 \ \boxed{} = 11$

Below, the same number goes into each box. Tell whether the statement is *true* or *false*.

7   If $25 - \boxed{} = 15$, then

    $15 + \boxed{} = 25$. **T F**

8   If $\boxed{} + 8 = 14$, then

    $14 - \boxed{} = 8$. **T F**

Functions

Here is a simple rule: Add 4. If you apply that rule to the numbers 0, 5, and 10, you get 4, 9, and 14, respectively. A rule that changes one value to another value is called a **function.**

PRACTICE

For each problem, tell the rule that works for all 3 boxes.

Sample

5 ☐ = 7   Add 2.
2 ☐ = 4   _____
1 ☐ = 3

1   4 ☐ = 9   _____
10 ☐ = 15
21 ☐ = 26

2   14 ☐ = 11   _____
10 ☐ = 7
23 ☐ = 20

3   1 ☐ = 12   _____
21 ☐ = 32
50 ☐ = 61

4   2 ☐ = 4   _____
3 ☐ = 6
4 ☐ = 8

For problems 5–7, tell what rule will change each "In" number to the corresponding "Out" number.

Sample

In	3	4	5	7
Out	1	2	3	5

Subtract 2.

5

In	1	3	0	4
Out	5	7	4	8

_____

6

In	12	8	18	16
Out	4	0	10	8

_____

7

In	4	12	8	10
Out	16	24	20	22

_____

Circle the letter of the rule for each pattern.

8

In	1	2	3	4
Out	4	6	8	10

A   Add 1, then multiply by 2.
B   Multiply by 5, then subtract 1.
C   Add 3.

9

In	1	2	3	4
Out	1	3	5	7

F   Add 0.
G   Add 2.
H   Multiply by 2, then subtract one.

Writing Letters and Symbols for Words

An algebra problem uses letters to take the place of numbers. These letters are called **variables** or **unknowns.** A letter can represent a single number or it can represent many values.

In	3	4	5	7
Out	1	2	3	5

In this table, suppose that x represents each "In" number. Then each "Out" number can be described as $x - 2$.

a number increased by five

In this example, if you use the letter n to represent "a number," this phrase can be described as $n + 5$.

PRACTICE

Write an algebra expression for each phrase. Let the letter n stand for the unknown number. *Hint:* If you have trouble, review page 38.

1   a number plus thirty-two   _____

2   a number times three   _____

3   the sum of a number and twenty-four   _____

4   six more than a number   _____

5   seven less than a number   _____

6   half of a number   _____

7   the total of a number and eight   _____

8   the product of a number and six   _____

9   thirteen divided by a number   _____

10   a number decreased by six   _____

11   a number multiplied by fourteen   _____

12   a number increased by ten   _____

13   two times a number   _____

14   five less than a number   _____

15   three-fourths of a number   _____

16   twice a number   _____

17   the difference between nine and a number   _____

18   a number over twelve   _____

19   a number times ten   _____

Solving word problems in an important use of the letters and symbols of algebra. The key steps are to use a letter for an unknown value and to look for signal words for addition, subtraction, multiplication, and division.

Write an algebra expression to describe each situation.

Sample
Jack earns w dollars an hour. Starting next month, he will get $2.00 more per hour. What will be his new hourly wage?
Solution: $w + 2$ dollars

20 Let r stand for the rent Max now pays each month. Starting next month, Max will pay 20 dollars more per month. What will be his new monthly rent payment?

_____

21 There are p pieces in an extra large pizza. If 4 friends evenly divide an extra large pizza, how many pieces will each friend get?

_____

22 Loraine makes x dollars a month. Each month $450 is taken out of her paycheck for taxes. What is Loraine's monthly take-home pay?

_____

23 Rock candy costs c cents a pound. What is the cost of 3 pounds of rock candy?

_____

24 In the talent show, there will be d dancers and m musicians. How many performers will there be in all?

_____

25 There will be k skaters in the competition. Each skater will perform for exactly 4 minutes. What is the total amount of time that the skaters will be performing?

_____

26 There were 28 students enrolled in Elise's class, but d students dropped out. How many students are in the class now?

_____

27 Karen does not have enough cups of flour for a bread recipe. She needs 6 cups of flour, but she has only F cups of flour. How many more cups of flour does Karen need?

_____

28 The last time Lynn saw Ricky, he was only 7 years old. Now he is y years old. How long has it been since Lynn saw Ricky?

_____

29 Maria weighed 172 pounds right before her baby was born. Afterward, she weighed x pounds. How much weight did she lose when her baby was born?

_____

Writing Letters and Symbols for Words

Writing Equations

To solve a word problem using algebra, you must write an **equation.** An equation is an expression such as **n + 7 = 10.** The equal sign in the equation says the same value is on each side of that sign.

Example 1
Armen is 48 years old. How old was he 5 years ago?

What operation should I use?
The phrase "5 years ago" is a signal for subtraction.
What do I perform the operation on?
Armen's present age (48) is the total, so you must subtract 5 years *from* that total.
Equation: 48 − 5 = x

Example 2
Terrence just bought 3 new wrenches. He now has a total of 15 wrenches. How many wrenches did Terrence have to begin with?

What operation should I use?
The word "total" is a signal for addition.
What do I perform the operation on?
The total is 15, so you must add the other two numbers together to get 15.
Equation: x + 3 = 15

PRACTICE

For each problem, write an equation that describes the problem. Use the letter x for the unknowns. (Hint: You subtract *from* a total. To *get* a total, you add.)

1   A month ago, Cecelia had 7 pairs of shoes. Now she has 12 pairs. How many pairs has she bought in the last month?   _____

2   Howard bought 5 big bags of potato chips. Now there are $1\frac{1}{2}$ bags left. How many bags of potato chips has his family eaten?   _____

3   Clarice worked at her first job for 3 years. Then she worked at Available Temps for 7 years. How many years did she work?   _____

4   Tricia is 15 years older than her cousin. When her cousin is 65 years of age, how old will Tricia be?   _____

5   Last month, Arlene sponsored a number of cats at the Humane Society. This month she increased the number of cats by 6, and now she sponsors a total of 15 cats. How many cats did she sponsor last month?   _____

6   Rochelle had 24 bags of used clothes. She gave 7 of the bags to a charity. How many bags are left?   _____

7   There are 12,000 people living in Bloomington. Ten years ago, the population was 9,000 people. How much has the population grown in the last ten years?   _____

Solving Equations

The equation $x + 3 = 15$ means "add an unknown number and 3; the result is 15." To solve that equation, perform the *inverse* (the opposite) of the operation shown. The inverse of "add 3" is "subtract 3." To solve the equation, subtract 3 from each side of the equation.

To solve an addition problem, you subtract.

Equation	$x + 3 = 15$
Subtract 3 from each side.	$\underline{ - 3 \quad -3}$
	$x + 0 \qquad 12$
Adding 0 to a number does not change its value, so	$x = 12$

To solve a subtraction problem, you add.

$$x - 3 = 6$$
$$\underline{+ 3 + 3}$$
$$x \quad\;\; = 9$$

Some problems need two steps.

$$15 - x = 7$$
$$\underline{+ x \qquad + x}$$
$$15 \qquad = 7 + x$$
$$\underline{-7 \qquad\;\; - 7}$$
$$8 \qquad = \quad x$$

PRACTICE

Solve each of the following equations. Then check your answer by seeing whether is makes the original equation true. *Remember:* **Always do the same thing to both sides of an equation.**

1. $x + 12 = 20$ $x =$ _____

2. $x + 44 = 69$ $x =$ _____

3. $9 + 6 = x$ $x =$ _____

4. $x - 13 = 12$ $x =$ _____

5. $45 - 6 = x$ $x =$ _____

6. $x + 14 = 54$ $x =$ _____

7. $x + 32 = 47$ $x =$ _____

8. $x + 6 = 19$ $x =$ _____

9. $x - 39 = 64$ $x =$ _____

10. $48 - x = 42$ $x =$ _____

Use algebra to solve these word problems. First, write an equation to describe each problem. (This is called *setting up the equation*.) Then solve your equation.

11 A number increased by seven is fifteen. What is the number?

equation: _____

solution: _____

12 When you subtract 6 from a number, the difference is 7. What is the number?

equation: _____

solution: _____

13 Jason spent $20 on theater tickets. He bought 4 tickets. How much did each ticket cost?

equation: _____

solution: _____

14 Maria is 12 years older than Anita. Anita is 35 years old. How old is Maria?

equation: _____

solution: _____

15 Cynthia has twice as many rose bushes as her sister. Cynthia has 28 rose bushes. How many rose bushes does her sister have?

equation: _____

solution: _____

16 Brad's total bill for lunch was $6.25. All he bought was a sandwich for $4.50 and a drink for $1.25. How much was the tax?

equation: _____

solution: _____

17 There are twice as many women as men in the choir. There are 8 men in the choir. Altogether, how many people are in the choir?

equation: _____

solution: _____

Algebra Skills Practice

Circle the letter for the correct answer to each problem.

1 What number goes in both boxes to make these number sentences true?

$$4 + \square = 16$$
$$16 - \square = 4$$

    A   4
    B   8
    C   13
    D   12

2 If $8 \times n = 48$, then n is ___?___ .

    F   40
    G   6
    H   4
    J   52

3 Tell what rule will change each "Input" number to the corresponding "Output" number?

In	3	2	5	8
Out	12	10	16	22

    A   Add 3, then multiply by 2.
    B   Multiply by 3, then add 3.
    C   Multiply by 4.
    D   Add 9.

4 What number is missing from this number pattern?

56, 50, 44, 38, 32,__?__, 20, 14

    F   26
    G   24
    H   22
    J   28

5 Which group of numbers is missing from this number pattern?

9, 12, 15,__?__,__?__,__?__, 27, 30, 33

    A   17, 19, 21
    B   17, 20, 23
    C   18, 21, 24
    D   18, 22, 24

6 Which number sentence *does not* have 5 as a solution?

    F $5 - 1 = \square$
    G $5 \div 1 = \square$
    H $5 \times 1 = \square$
    J $5 - 0 = \square$

7 Which operation sign goes in the box to make the number sentence true?

$$9 \; \square \; 9 = 1$$

    A   +
    B   ÷
    C   −
    D   ×

8 How many squares will be in the next entry in the pattern below?

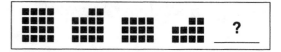

    F   6
    G   8
    H   10
    J   12

9 What number is missing from this number sequence?

   _?_ , 21, 24, 27, 30, 33

   A 16
   B 17
   C 18
   D 19

10 If $80 \div x = 8$, what is the value of x?

   F 10
   G 72
   H 11
   J 5

11 Here is a pattern:

   X X 0 X X 0 X X 0 X X 0

   Which of these describes the pattern?

   A one X, then one 0
   B two X's, then two 0's
   C two X's, then one 0
   D one X, then two 0's

12 Here is a number pattern:

   1, 5, 9, 13, 17, 21, 25

   Which of these describes the number pattern?

   F Add 4 to the last number.
   G Add 3 to the last number.
   H Count by fives.
   J Multiply the last number by 5.

13 What sign goes in the box to make this number sentence true?

   6 ☐ 3 = 2

   A +
   B −
   C ×
   D ÷

14 This list shows the wages given to laborers at a construction company.

Experience	Hourly Wage
None	$5.10
1 year	$5.40
2 years	$5.70
3 years	$6.00
4 years	$6.30

What pattern, if any, is shown in the list?

   F Wages double with every year of experience.
   G Wages increase 30 cents with every year of experience.
   H The more experience a laborer has, the larger his or her raises.
   J There is no pattern in the table.

15 Louella got a $900 bonus at work. She put $500 in savings and spent half of the rest on a business suit. How much did she spend on the business suit?

   A $500
   B $400
   C $200
   D $100

Measurement

Choosing the Best Tool

An important part of measuring is to pick the correct tool. For example, to measure the weight of a car you would use a truck scale. To measure the weight of a computer disk you could use a postal scale.

PRACTICE

Tell whether each tool below measures weight, capacity, time, length, or temperature.

1 ruler _____

2 doctor's scale _____

3 glass measuring cup _____

4 measuring spoons _____

5 tape measure _____

6 car odometer (mileage gauge) _____

7 clock _____

8 calendar _____

9 thermometer _____

10 truck scale _____

For each problem below, write the letter of the *best* tool for measuring the object.

A glass measuring cup
B tape measure
C small postal scale
D odometer
E ruler
F measuring spoons
G bathroom scale

11 the height of a wall _____

12 the width of a coffee cup _____

13 the distance between two towns _____

14 the width of a sheet of paper _____

15 your weight _____

16 the weight of a letter _____

17 enough coffee grounds to brew one cup _____

18 the amount of liquid in a pitcher _____

Reading a Scale

A number line on a measurement tool is called a **scale.** To use a simple scale, like the ruler shown below, follow these steps:

STEP 1. Line up the zero with one end of the object you are measuring.
STEP 2. Find the number that lines up with the other end of the object.
STEP 3. Check your work by measuring the object again.

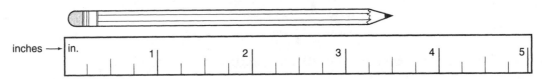

Notice that there are small lines, or **tick marks,** between the numbers on this scale. Tick marks divide the distance between numbers into equal-sized parts.

To find out what each tick mark represents, count the number of spaces that are formed between each pair of numbers on the scale. If two spaces are formed between neighboring numbers, than each tick mark represents half the value of the difference of the two numbers. If four spaces are formed, each tick mark is one-fourth of the value, and so on.

On the ruler above, the inches are divided into four parts, so each tick mark represents one-fourth of an inch. The end of the pencil lines up with the second tick after 3 inches, or $3\frac{2}{4}$ inches. The pencil is $3\frac{1}{2}$ inches long.

PRACTICE

Use a ruler to find *six* of the measurements listed below. (You can use the ruler shown above.) Give your answers to the nearest $\frac{1}{4}$ inch.

1 length of a C battery _____ inches

2 length of a D battery _____ inches

3 height of a standard playing card _____ inches

4 width of a standard playing card _____ inches

5 height of a check book _____ inches

6 width of a half-pint milk carton _____ inches

7 height of a *TV Guide*® _____ inches

8 width of a *TV Guide*® _____ inches

9 height of a roll of toilet paper _____ inches

10 thickness of a video tape _____ inches

Reading a Scale That Skips Numbers

Sometimes a scale skips numbers. For instance, thermometers and bathroom scales are usually labeled in tens: 10, 20, 30, 40, and so on. On such scales, the tick marks stand for whole numbers, not fractions.

Below, the pointer is about halfway between 10 and 20.

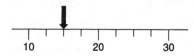

Since 15 is halfway between 10 and 20, the reading is about 15.

PRACTICE

Fill in each blank below.

1 The pointer on a scale is about halfway between 12 and 14. What number is halfway between 12 and 14? _____

2 The pointer on a scale is about halfway between 40 and 50. What number is halfway between 40 and 50? _____

3 The pointer on a scale is about halfway between 0 and 20. What number is halfway between 0 and 20? _____

4 The pointer on a scale is $\frac{1}{4}$ of the way from 0 to 20. What number is $\frac{1}{4}$ of the way from 0 to 20? _____

5 The pointer on a scale is about $\frac{3}{4}$ of the way from 0 to 20. What number is $\frac{3}{4}$ of the way from 0 to 20? _____

6 The pointer on a scale is about halfway between 60 and 70. What number is halfway between 60 and 70? _____

7 What is the reading on this scale? _____

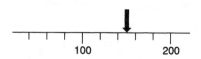

8 What is the reading on this scale? _____

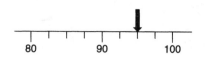

When a scale skips numbers, follow these steps to figure out what the tick marks represent.

STEP 1. Find the difference between two neighboring numbers on the scale.
STEP 2. Count the number of spaces formed by the tick marks between the two numbers.
STEP 3. Divide the difference between the numbers by the number of spaces.

In the scale below, there are 5 tick marks between 50 and 100, so each tick mark represents 10 units.

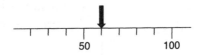

The arrow is 1 tick mark beyond 50, so the reading is 50 + 10, or 60 units.

PRACTICE

Read each measurement by using the tick marks.

9   On a scale, the area between 10 and 20 is divided into 5 spaces. How many units is each tick mark worth? _____

10 On a scale, the area between 20 and 40 is divided into 10 spaces. What is each tick mark worth? _____

11 On a scale, the area between 60 and 80 is divided into 5 spaces. What is each tick mark worth? _____

12 On a scale, the area between 25 and 50 is divided into 5 spaces. What is each tick mark worth? _____

13 On a scale, the area between 0 and 50 is divided into 5 spaces. What is each tick mark worth? _____

14 Below, each tick mark is worth _?_ units, and the reading on the scale is _?_ . _____

15 Below, each tick mark is worth _?_ units, and the reading on the scale is _?_ . _____

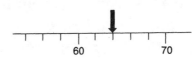

Measuring Temperature

Temperature is measured using a **thermometer** marked in **degrees Fahrenheit** (°F) or **degrees Celsius** (°C). Look for one of those two symbols on a thermometer to tell you which units are used.

PRACTICE

Use this thermometer for questions 1–4.

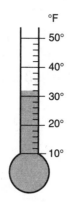

1 Circle one: Which unit does this thermometer use?

   degrees Celsius   degrees Fahrenheit

2 What temperature is shown on the thermometer?

   _____

3 Is 20°F colder or warmer than the temperature shown above?

   _____

4 Water freezes at the temperature shown above. Which of these would be the best setting for your freezer?

   A 40°F   **C** 56°F
   B 35°F   **D** 5°F

This thermometer shows both degrees Celsius and degrees Fahrenheit. Use it for questions 5 through 8.

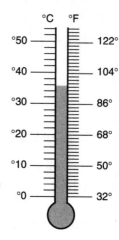

5 Circle the larger unit:

   one degree Celsius

   one degree Fahrenheit

6 Each tick mark on these scales represents how many degrees?

   _____ °C   _____ °F

7 What temperature is shown on the thermometer?

   about _____ °F

   about _____ °C

8 Which of these measurements is the hottest temperature?

   A 104°F   **C** 40°C
   B 50°C   **D** 86°F

Measuring Length

The distance between two points can be a measurement of *width, height,* or *depth.* Each of these measurements is a **length.** The basic **standard units of length** are **inch, foot, yard,** and **mile.**

Standard Units of Length

12 inches (in.) = 1 foot (ft) 3 feet = 1 yard (yd) 5,280 feet = 1 mile (mi)	An inch is about the length of a straight pin. A foot is about the length of a man's foot. A yard is about the length of your arm. A mile is about 10 city blocks.

PRACTICE

Circle the best estimate for each measurement.

1   the length of a room

    A $2\frac{1}{2}$ feet

    B 36 inches

    C 15 feet

2   the length (long side) of a video tape

    F $\frac{1}{2}$ foot

    G 1 yard

    H 3 inches

3   the width of your thumbnail

    A 3 inches

    B $\frac{1}{2}$ inch

    C 6 inches

4   the height of a garbage can

    F 3 yards

    G 12 inches

    H 3 feet

5   the width of a car

    A 6 feet

    B 12 feet

    C 7 yards

Use this information to fill in the blanks below.

To change feet into inches, multiply by 12.
To change inches into feet, divide by 12.
To change feet into yards, divide by 3.
To change yards into feet, multiply by 3.

6   A fence post is 2 feet tall. How many inches is that?   _____

7   A desk is 1 yard tall. How many feet is that?   _____

8   How many inches are in half a foot?   _____

9   A cabinet is 6 feet tall. How many yards is that?   _____

10   36 inches = _?_ feet   _____

11   $1\frac{1}{2}$ feet = _?_ inches   _____

12   5 yards = _?_ feet   _____

13   9 feet = _?_ yards   _____

Using a Ruler

Most standard rulers, yardsticks, and tape measures are not as simple as the scales you used on the previous pages. They use several different types of tick marks:

You will see the following on most rulers marked in inches and feet:

◆ The longest tick marks divide the ruler into inches.
◆ The next longest tick marks divide each inch into half-inches.
◆ The half inches are divided into fourths by slightly shorter tick marks.
◆ The fourths are divided into eighths by even shorter tick marks.
◆ The shortest tick marks may be eighths or sixteenths.

PRACTICE

Find a ruler marked in inches. Look at the long tick mark halfway between each pair of numbers. Use those long tick marks to measure each line below to the nearest $\frac{1}{2}$ inch. Some answers will be whole numbers.

1 ———————————————————

_____ inches

2 ————————————

_____ inches

3 —————————

_____ inches

4 ——————

_____ inches

5 ———————————————

_____ inches

Find the tick marks on your ruler that divide each inch into four equal parts, or fourths. Use those marks to measure the lines below to the nearest $\frac{1}{4}$ inch.

6 ————

_____ inches

7 ———————————

_____ inches

8 —————————

_____ inches

9 ——

_____ inches

10 ———————————————

_____ inches

Find the height and the width of *four* objects listed below. Give your answers to the nearest $\frac{1}{4}$ inch.

11   2-liter soda bottle   _____

12   business card   _____

13   Time® Magazine   _____

14   a piece of notebook
     paper   _____

15   a business envelope   _____

16   a stick of butter   _____

17   a record album cover   _____

18   this book   _____

Find the marks on your ruler that divide inches into eighths. Count the eighth-inch tick marks between 1 and 2 inches to make sure you are looking at your ruler correctly.

Measure each line below to the nearest eighth of an inch.

19   ———————————————

_____ inches

20   ——

_____ inches

21   ———

_____ inches

22   —————————

_____ inches

Find each measurement listed below. Give your answers to the nearest sixteenth of an inch. You can use the following facts:

$$\frac{1}{2} = \frac{8}{16} \qquad \frac{1}{4} = \frac{4}{16} \qquad \frac{1}{8} = \frac{2}{16}$$

23   width of a credit
     card   _____ inches

24   of a dollar bill   _____ inches

25   width of a postage
     stamp   _____ inches

26   width of a quarter
     (Measure it at its
     widest point.)   _____ inches

27   width of a dime   _____ inches

Metric Units of Length

The **metric units of length** include **millimeters, centimeters, meters,** and **kilometers.** The ruler below shows two different scales. One side is marked in inches, and the other side is marked in centimeters.

Metric Units of Length
10 millimeters (mm) = 1 centimeter (cm)
1 meter (m) = 100 centimeters
1 kilometer (km) = 1,000 meters

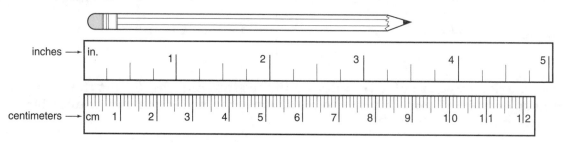

PRACTICE

Use the ruler above to answer the following questions.

1 About how many centimeters are shown on this ruler? _____

2 About how many inches are shown on the ruler? _____

3 About how many centimeters are in one inch? (Estimate to the nearest $\frac{1}{2}$-cm.) _____

4 In centimeters, about how long is the pencil? _____

5 In millimeters, about how long is the pencil? (*Hint:* You can multiply the number of centimeters by 10.) _____

6 Suppose that an object is 4 inches long. About how long is that object in centimeters? (Estimate to the nearest $\frac{1}{2}$-centimeter.) _____

In the metric system, you can multiply or divide by powers of 10 to change between units. (The **powers of 10** are 10, 100, 1,000, and so on.)

◆ Multiply by 10 to change centimeters to millimeters. Multiply by 100 to change meters to centimeters. Multiply by 1,000 to change kilometers to meters or to change meters to millimeters.

◆ Divide by 10 to change millimeters into centimeters. Divide by 100 to change centimeters to meters. Divide by 1,000 to change meters to kilometers or to change millimeters to meters.

7 How many millimeters are in 7 centimeters? _____

8 How many centimeters are in 5 meters? _____

9 300 meters = _?_ kilometers _____

Finding Perimeter

The total length of the sides of a figure is called the **perimeter.**

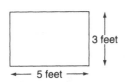

For this figure, the top edge and the bottom edge have the same length. Each is 5 feet. The left edge and right edge is each 3 feet. The perimeter is
$$5 + 3 + 5 + 3 = 16 \text{ feet}$$

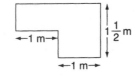

For this figure, the top edge is 1 meter + 1 meter = 2 meters. The right edge is $1\frac{1}{2}$ meters, so the sum of the two parts of the left edge is $1\frac{1}{2}$ meters. The perimeter is
$$2 + 1\frac{1}{2} + 2 + 1\frac{1}{2} = 7 \text{ meters}$$

We do not know the lengths of the two unmarked sides. So we do not have enough information to calculate the perimeter of the figure.

PRACTICE

Find the perimeter of each shape. *Hint:* **The symbol " is an abbreviation for inches and the symbol ' is an abbreviation for feet.**

1   a rectangular room with sides of 12 feet, 5 feet, 12 feet, and 5 feet

perimeter: _____

2

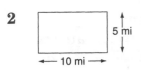

perimeter: _____

3

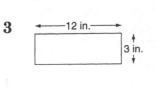

perimeter: _____

4

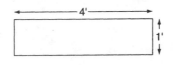

perimeter: _____

5

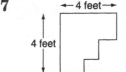

a square

perimeter: _____

6

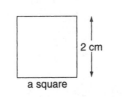

perimeter: _____

7

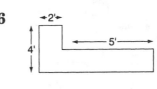

perimeter: _____

8   The perimeter of a rectangle is 20 cm. The top edge and bottom edge are each 8 cm long. How long is each of the right and left edges?

_____

Finding Area

The amount of a flat region inside a figure is called its **area.** To find the area of a square or a rectangle, you can multiply length times width.

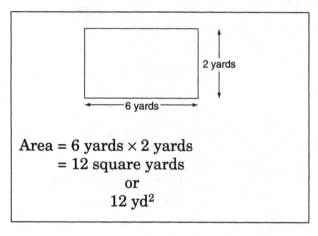

Area = 6 yards × 2 yards
= 12 square yards
or
12 yd²

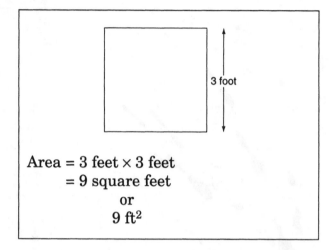

Area = 3 feet × 3 feet
= 9 square feet
or
9 ft²

The abbreviation "yd²" is pronounced "square yards," and the abbreviation "ft²" is pronounced "square feet." Area is always measured in *square units*.

PRACTICE

Find the area of each shape below. Be sure to give your answers in square units.

1 a rectangular tile that is 6 centimeters long and 6 centimeters wide

Area: _____

2 a rectangle whose sides measure 2 mm, 4 mm, 2 mm, and 4 mm

Area: _____

3

5 mi

10 mi

Area: _____

4

12 inches

3 inches

Area: _____

5

4 ft

1 ft

Area: _____

6

2 cm

a square

Area: _____

7

3 m

3 m

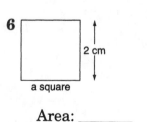

Area: _____
(*Hint:* The area is <u>not</u> that of a 6 m × 6 m square.)

8 The area of a square is 25 square inches. How long is each side of the square?

_____

Measuring Weight

For weight, the **standard units** are **ounces, pounds,** and **tons.** The **metric units** for weight are **milligrams, grams,** and **kilograms.**

Units of Weight

1 pound (lb) = 16 ounces (oz) 1 ton (T) = 2,000 lb	A pencil weighs about 1 ounce. An eggplant weighs about 1 pound. A car weighs about 1 ton.
1 gram (g) = 1,000 milligrams (mg) 1 kilogram (kg) = 1,000 grams (g)	A needle weighs about 1 milligram. A peanut weighs about 1 gram. A telephone book weighs about 1 kilogram.

PRACTICE

Use the table above for questions 1–6.

1 How many ounces are in half a pound? _____

2 How many ounces are in $1\frac{1}{2}$ pounds? _____

3 How many grams are in half a kilogram? _____

4 How many milligrams are in one quarter of a gram? _____

5 Which of these would weigh about one ounce?

   A a houseplant   **C** a bike
   B a quarter   **D** a lamp

6 Which of these would weigh about one pound?

   F a computer disk
   G a dinner roll
   H a garbage can
   J a loaf of bread

Use the information below for questions 7–12.

To change:
  ounces into pounds, divide by 16.
  pounds into ounces, multiply by 16.
  grams into kilograms, divide by 1,000.
  kilograms into grams, multiply by 1,000.

7 A package weighs 1 pound, 3 ounces. How many ounces is that? _____

8 A supermarket chicken weighs 2 kg, 550 g. How many kilograms is that? _____

9 A box of biscuit mix weighs $\frac{1}{4}$ kilogram. How many grams is that? _____

Circle the heavier weight in each pair.

10   3 grams   1,200 milligrams

11   900 grams $\frac{1}{2}$ kilogram

12   1 pound   15 ounces

Measuring Liquid Capacity

For liquids, the standard units are **cups, pints, quarts,** and **gallons.** The metric units are **liters** and **milliliters.**

Units of Capacity

1 pint = 2 cups	An ice cream dish holds 1 cup.
	A mug holds about 1 pint.
1 quart = 4 cups	A thin milk carton holds 1 quart.
(or 2 pints)	
1 gallon = 4 quarts	A plastic milk jug holds 1 gallon.
(or 8 pints)	
1 liter = 1,000 ml	A plastic soda bottle holds 2 liters.

To change a measurement into larger units, you can divide. **Examples:** To change milliliters into liters, divide by 1,000. To change cups into quarts, divide by 4.

To change a measurement into smaller units, you can multiply. **Examples:** To change liters into milliliters, multiply by 1,000. To change quarts into cups, multiply by 4.

PRACTICE

Use the tables above to answer these questions.

1 A juice box holds about __?__ .

    A 1 cup   **C** 1 gallon
    B 1 quart   **D** 1 liter

2 A full bathtub holds about __?__ .

    F 50 cups   **H** 50 pints
    G 50 gallons **J** 50 milliliters

3 A few drops is about 5 __?__ .

    A pints   **C** gallons
    B quarts   **D** milliliters

4 A child's pail of sand holds about __?__ .

    F 1 gallon   **H** 1 liter
    G 5 quarts   **J** 7 gallons

5 Which is larger, a gallon or a liter? _____

6 Which is larger, 3 pints or 1 gallon? _____

7 Which is larger, 1 gallon or 6 liters? _____

8 3 pints = __?__ cups _____

9 3 quarts = __?__ cups _____

10 $\frac{1}{2}$ gallon = __?__ cups _____

11 32 cups = __?__ gallons _____

12 $1\frac{1}{2}$ gallons = __?__ cups _____

13 $\frac{1}{4}$ gallon = __?__ cups _____

14 5 quarts = __?__ gallons _____

Measuring Time

The clock at the right shows the time: forty-three minutes after four o'clock. Using figures, that time is written **4:43,** just as it would appear on a digital clock. The number to the left of the colon is the hour. The number to the right of the colon represents minutes. When you say a time, you say the hour and then the minutes.

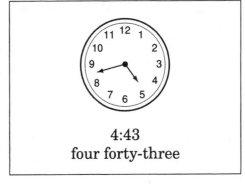

4:43
four forty-three

Units of Time

1 minute (min) = 60 seconds (sec)

1 hour (hr) = 60 min

1 day = 24 hr

1 week = 7 days

1 month = about $4\frac{1}{2}$ weeks

1 year (yr) = 12 months or 52 weeks

To add two time durations, add the hours and add the minutes. If the sum of the minutes is more than 59, convert the extra minutes to hours.

```
  1 hour, 15 minutes
+ 2 hours, 50 minutes
  3 hours, 65 minutes
```

Simplify:
65 minutes = 1 hour, 5 minutes
So 3 hours, 65 minutes = 4 hours, 5 minutes.

PRACTICE

Use the table above to convert these time durations.

1   $\frac{1}{2}$ hour = _?_ minutes   _____

2   3 hours = _?_ minutes   _____

3   120 minutes = _?_ hours   _____

4   2 weeks = _?_ days   _____

5   2 days = _?_ hours   _____

6   $\frac{1}{4}$ hour = _?_ minutes   _____

7   $\frac{1}{3}$ hour = _?_ minutes   _____

8   $1\frac{1}{2}$ hours = _?_ minutes   _____

9   $\frac{1}{2}$ day = _?_ hours   _____

Add the time durations below. If the sum of the minutes is 60 or more, convert the extra minutes to hours.

10   1 hour, 15 minutes
      + 4 hours, 30 minutes

11   45 minutes
      + 45 minutes

12   3 hours, 35 minutes
      + 2 hours, 45 minutes

13   5 hours, 10 minutes
      + 3 hours, 50 minutes

14   1 hour, 15 minutes
        3 hours, 39 minutes
      +   30 minutes

Changes in Time

For the problem below, imagine how the hands on the clock move. Count the number of *complete* hours that pass. Then count the minutes that pass.

For the problem below, you want to find when you should start a task so you will finish at a particular time. To do this, count the change in hours first. Then add the change in minutes.

Problem
Eric started work at 8:25. He finished at 4:10. How long did he work?

Count the hours: The number of *complete* hours from 8:25 to 3:25 is 7.
Count the minutes:
From 3:25 to 4:10 is 45 minutes.
Answer: 7 hours, 45 minutes

Problem
It takes 1 hour, 15 minutes to drive to the doctor's office. You have a 10:00 appointment. When should you leave?

You need a time *before* 10:00, so count back from 10:00.
Count the hours:
1 hour before 10:00 is 9:00.
Count the minutes:
15 minutes before 9:00 is 8:45.
Answer: You should leave at 8:45.

PRACTICE

Fill in the blank for each problem.

1   What time is 5 hours before 2:00?

2   It will take 6 hours to make bread. You plan to eat at 5:00. By what time should you start making the bread?

3   It will take 50 minutes to drive to your mother's house. You leave at 3:30. When will you arrive?

4   You began working at 9:15 A.M. It is now 1:35 P.M. How long have you been working?

5   You promised to spend $5\frac{1}{2}$ hours at a shelter. If you start at 8:15 A.M., when will you finish?

Measurement Skills Practice

Circle the letter for the correct answer to each problem.

1 What time will the clock show in 20 minutes?

  A 2: 20   **C** 3:20
  B 2:55   **D** 3:05

2 A pork chop weighs 8 ounces. What fraction of a pound is that?

  F $\frac{1}{4}$ **H** $\frac{1}{2}$

  G $\frac{1}{3}$ **J** $\frac{1}{5}$

3 Which of these instruments can be used to measure length?

  A a bathroom scale
  B a ruler
  C a telescope
  D a clock

4 You want to make a $\frac{1}{4}$-pound hamburger. How many ounces of hamburger do you need?

  F 10   **H** 4
  G 25   **J** 40

5 Maya leaves her children with the babysitter at 6:50 P.M. She returns at 8:45 P.M. How long were her children with the babysitter?

  A 1 hour, 55 minutes
  B 1 hour, 5 minutes
  C 2 hours, 5 minutes
  D 3 hours, 5 minutes

This diagram shows plans for a picture frame. Study the diagram. Then do Numbers 6 through 8.

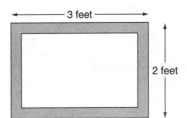

6 How many feet of framing would you need to form the border around the outside of this picture frame?

  F 5 feet   **H** 10 feet
  G 6 feet   **J** 8 feet

7 How much area will this picture and frame cover on the wall?

  A 6 square feet
  B 5 square feet
  C 10 square feet
  D 8 square feet

8 It takes about 25 minutes to cut the materials for this frame. It takes another 45 minutes to put them together. How long does it take to make the frame?

  F 1 hour, 10 minutes
  G 1 hour, 7 minutes
  H 1 hour, 20 minutes
  J $1\frac{1}{2}$ hours

9 What temperature is shown on this thermometer?

  A 65°C
  B 63°C
  C $1\frac{1}{2}$°C
  D 66°C

10 Which of the following would be the best unit to use to weigh a business letter?

    F pounds
    G tons
    H kilograms
    J grams

11 Byung wants to spend 20 minutes exercising on a stair machine. He starts at 4:52. When will he finish?

    A 5:12
    B 4:72
    C 5:02
    D 5:07

12 Rochelle buys 20 feet of edging for her garden. She plans to use all 20 feet to outline one square flower bed. How long will each side of the flower bed be?

    F 10 feet
    G 5 feet
    H 4 feet
    J 3 feet

13 You want to buy 8 ounces of peas. They cost $1.00 per pound. How much will you pay?

    A $1.00
    B $2.00
    C 50 cents
    D 80 cents

14 On a commuter train, the trip from Forest Hills to Central Station takes 40 minutes. Continuing on the train, the trip from Central Station to the zoo takes another 50 minutes. How long will it take to travel from Forest Hills all the way to the zoo?

    F 1 hour, 10 minutes
    G 1 hour, 20 minutes
    H 1 hour, 30 minutes
    J 1 hour, 40 minutes

15 Which of these is the best tool for finding out whether or not a refrigerator will fit through a doorway?

    A a calendar
    B a measuring cup
    C a tape measure
    D a bathroom scale

16 Which of the following would be a comfortable temperature for your bath water?

    F 20 degrees Fahrenheit
    G 50 degrees Fahrenheit
    H 100 degrees Fahrenheit
    J 200 degrees Fahrenheit

17 Here are the dimensions of three rectangular fields.

    Field A: 50 feet by 20 feet
    Field B: 30 feet by 30 feet
    Field C: 30 feet by 40 feet

Which of the fields has the most area?

    A Field A
    B Field B
    C Field C
    D They all have the same area.

Geometry

Using Logic

When you find or know a pattern, you can use that pattern. When a pattern has several steps, a key to following the pattern is to be careful.

Solve these problems step-by-step. Be sure your thinking is correct before you move on to the next step. Make notes and pictures to organize your thoughts.

PRACTICE

Solve each problem below. *Hint:* First, cross out all the answers that *could not be correct.*

1

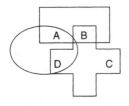

Which letter is inside exactly one figure?

_____

2

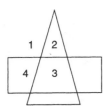

Which number is inside the rectangle but outside the triangle?

_____

3

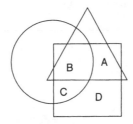

Which letter is inside the circle and the square, but outside the triangle?

_____

4

Which figure is a circle inside a square?

_____

5

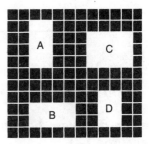

Which figure is 4 units (or squares) wide and 2 units tall?

_____

6

Which number is greater than 50 and is inside the circle?

_____

Points, Lines, and Angles

Geometry is the study of shapes. The ideas below are the building blocks used to create and describe shapes.

Basic Ideas in Geometry

Name	Meaning	Diagram	Symbol
point	a dot; a single location	•A	A, B, P, X, and so on
line	a straight, connected set of points; it extends in two directions	P Q	two letters with a double arrow above, such as $\overleftrightarrow{PQ}$
line segment	the part of a line that is between two endpoints	P Q	two letters with a bar above, such as $\overline{PQ}$
angle	an amount of a turn	A B C	the symbol $\angle$ followed by three letters, such as $\angle ABC$

PRACTICE

Use the diagram below to answer questions 1–3.

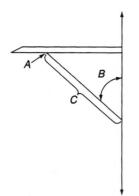

1 Which letter labels a point? _____

2 Which letter labels an angle? _____

3 Which letter labels a line segment? _____

4 $\angle FGH$ is the symbol for a(n) __?__ labeled with the points F, G, and H. _____

Use the diagram below to answer questions 5–7.

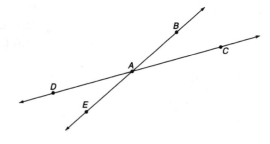

5 List all the points marked on this diagram.

_____

6 How many lines are shown in the diagram?

_____

7 Name three points that lie on the same line.

_____

Angles

The **size** of an angle refers to the opening between the sides of the angle. The point where the sides meet is called the **vertex** of the angle.

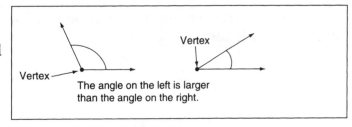

Vertex

Vertex

The angle on the left is larger than the angle on the right.

You can think of an angle as part of a circle. A complete circle is 360 degrees (or 360°). At the right, look at the angle made by a square corner with its vertex at the center of the circle. That angle represents $\frac{1}{4}$ of a circle, so its measurement is $\frac{1}{4}$ of 360° or 90°.

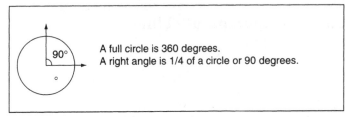

90°

A full circle is 360 degrees.
A right angle is 1/4 of a circle or 90 degrees.

PRACTICE

Below, circle the letter for the smaller angle in each pair.

1

 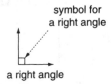

A   B

2

A   B

3

A   B

4

A   B

There is a special name for a square corner or 90 ° angle. It is called a **right angle**.

symbol for a right angle

a right angle

Circle the number(s) for the objects below that usually form right angles.

5   two walls meeting

6   two streets intersecting

7   the hands of a clock at 3:10

8   the back and the seat of a chair

9   the pages of an open book

10   the roof and floor of a building

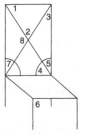

Use this diagram for questions 11, 12, and 13.

11   List all the angles marked on this chair that are right angles.

_____

12   List all the angles marked on this chair that are greater than 90°.

_____

13   List all the angles marked on this chair that are less than 90°.

_____

Lines

Two lines that cross or that will cross are called **intersecting lines.**

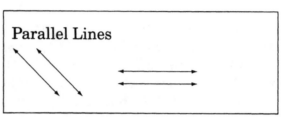

Intersecting Lines

Two lines that are always the same distance apart are called **parallel lines.**

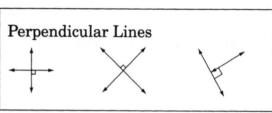

Parallel Lines

Two lines that form right angles when they meet are called **perpendicular lines.**

Perpendicular Lines

PRACTICE

Use the figures below to answer questions 1–3.

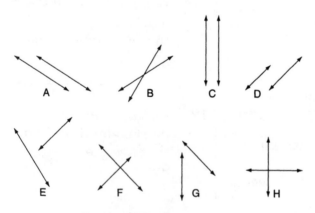

1   Which letters label pairs of lines that intersect? _____

2   Which letters label pairs of lines that are parallel? _____

3   Which letters label pairs of perpendicular lines? _____

Use this map for questions 4 and 5.

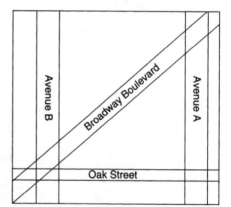

4   Which streets intersect Broadway Boulevard?

   _____

   _____

5   Which two streets are parallel?

   _____

   _____

Polygons

A **polygon** has straight sides and is named for its number of sides.

Polygons

Name	Shape	Sides
triangle		3
quadrilateral		4
pentagon		5
hexagon		6

Special Types of Polygons

Name	Examples	Meaning
regular polygon		In a regular polygon, all the sides are the same length.
parallelogram		A parallelogram has four sides. Opposite sides are parallel.
rectangle		A rectangle has four sides. It has four right angles.
square		A square has four sides. It has four right angles and four equal sides.

PRACTICE

Use this group of figures to answer the questions on this page.

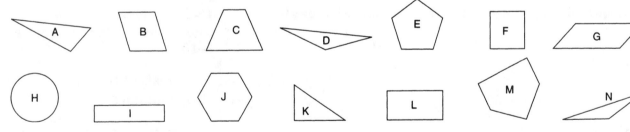

1   Which four figures are triangles?

_____

2   Which six figures are quadrilaterals?

_____

3   Which two figures are pentagons?

_____

4   Which figure is a hexagon?

_____

5   Which figure is a square?

_____

6   Which three polygons are regular?

_____

7   Which five figures are parallelograms?

_____

8   Circle all the terms below that describe a square.

regular   parallelogram
quadrilateral   pentagon

Triangles

Triangles are a particularly interesting and important type of figure. For the sides, the sum of any two sides must be greater than the third side. For the angles, the sum of the 3 angles is always 180°.

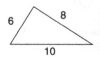

6 + 8 is greater than 10.
6 + 10 is greater than 8.
8 + 10 is greater than 6.

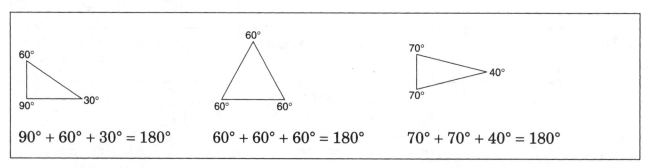

90° + 60° + 30° = 180°   60° + 60° + 60° = 180°   70° + 70° + 40° = 180°

This is a right triangle. It contains one 90-degree angle.

This is a regular triangle. All sides are equal and all angles are equal. A regular regular triangle is called an **equilateral** triangle.

In this triangle, two sides are equal and the angles opposite those sides are equal. This triangle is called an **isosceles** triangle.

PRACTICE

For questions 1–7, find the size of each third angle.

1
45°
90°

2
75°
75°

3
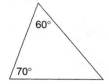
60°
70°

4

30°
40°

5
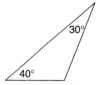
90°
42°

6
45°
105°

7
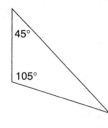
60°
60°

Use the triangles from questions 1–7 for questions 8–10.

8 List all the right triangles.

_____

_____

9 List all the equilateral triangles.

_____

_____

10 List all the isosceles triangles.

_____

_____

Circles

Another important shape is the circle. A **circle** is all the points that are a particular distance from a point called the **center** of the circle. The distance from the center to any point on the circle is called the **radius** of the circle. The width of the circle is called the **diameter,** and it is twice as long as the radius.

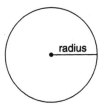

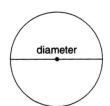

The diameter of a circle passes through its center point.

PRACTICE

1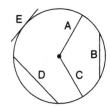

Which line segments in this diagram show the radius of the circle?

_____

2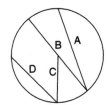

Which line segment represents the diameter of the circle?

_____

Find the *diameter* of each circle below. If you do not have enough information to find the diameter, write "cannot tell."

3

2 in.

diameter: _____

4

3 cm

diameter: _____

5

3 cm

diameter: _____

Find the *radius* of each circle below. If you do not have enough information to find the radius, write "cannot tell."

6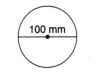

100 mm

radius: _____

7

4 cm

radius: _____

8

3 cm

radius: _____

Drawing Figures

An important skill in geometry is to draw or imagine figures. Also, you need to be able to look at figures from different points of view.

PRACTICE

For each problem on this page, draw or imagine a figure. Then circle the letter for the correct answer to each problem.

1   This picture shows a piece of paper that has been folded in half and then cut as shown. Once it is unfolded, what is the shape of the cut-out?

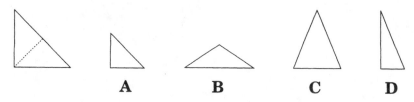

2   If you trace the bottom of this cylinder, what shape do you draw?

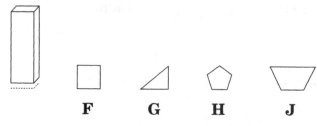

3   Below, what is the shape of the cut-out?

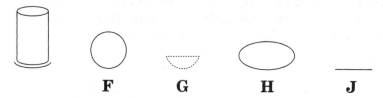

4   If you trace the bottom of this object, what shape do you draw?

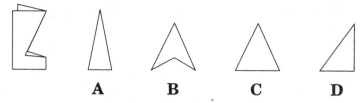

5   If you fold this figure along the dotted line, what is the shape of the *folded* figure?

Symmetry

If a figure can be divided into two identical halves, then we say that the figure has **symmetry** or it is **symmetric.** The line that separates the two parts is called the **line of symmetry.**

PRACTICE

For each problem, circle the letter for the figure in each row that has been divided into symmetric halves.

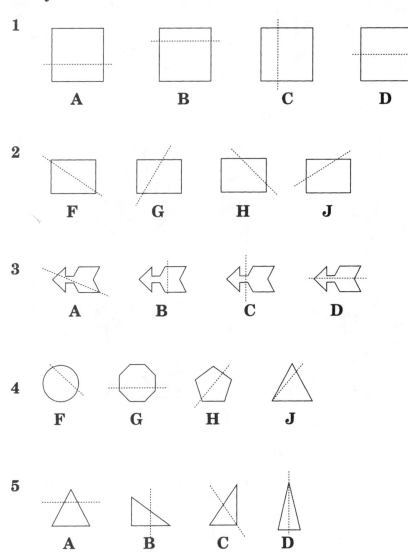

1

   A   B   C   D

2

   F   G   H   J

3

   A   B   C   D

4

   F   G   H   J

5

   A   B   C   D

Congruent Figures

If two figures have the same size and the same shape, they are called **congruent** figures.

PRACTICE

1 Which figure is congruent to the figure in the dark box?

 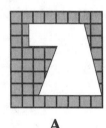  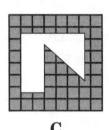

A   **B**   **C**

2 Which figure is congruent to the figure in the dark box?

 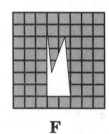 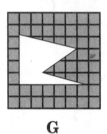 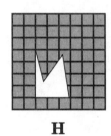

F   **G**   **H**

3 Which figure is congruent to the figure in the dark box?

 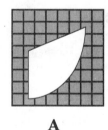 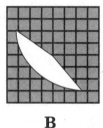 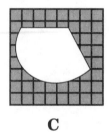

A   **B**   **C**

Use the diagram below for problems 4 through 6.

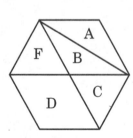

4 Which figure is congruent to figure *A?*

_____

5 Which figure is congruent to figure *C*?

_____

6 Figure *D* is congruent to the shape formed by joining figures _?_ and _?_ .

_____

Similar Figures

If two figures have the same shape, but not necessarily the same size, they are **similar** figures.

PRACTICE

1 Which figure is similar to the dark figure?

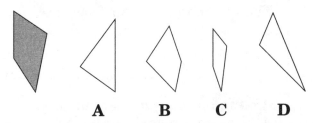

A   **B**   **C**   **D**

2 Which figure is similar to the dark figure?

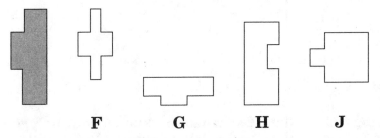

F   **G**   **H**   **J**

3 Which figure is similar to the dark figure?

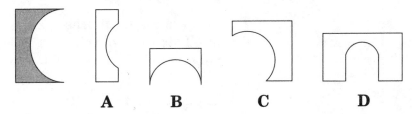

A   **B**   **C**   **D**

Use the diagram below for problems 4 and 5.

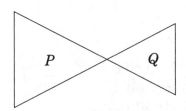

4 Do triangles P and Q look similar?

_____

5 Do triangles P and Q look congruent?

_____

Use the diagram below for problems 6 and 7.

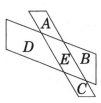

6 Which figure, if any, looks similar to figure A?

_____

7 Which figure, if any, looks similar to figure B?

_____

Three-Dimensional Figures

A **three-dimensional figure** has length, width, and depth. Another name for a three-dimensional figure is **solid figure**.

Common Solid Figures

A **cube** has six sides, and each side is a square. Every corner of a cube forms a right angle.

A **cylinder** is shaped like a can. The top and bottom surfaces are circles.

A **rectangular solid** is the shape of a box. It has six sides, and each side is a rectangle. Every corner of a rectangular solid is a right angle.

A **cone** has one circular surface. At the opposite end, the cone comes to a point.

PRACTICE

Tell whether each of the following objects is shaped like a cube, a rectangular solid, a cylinder, or a cone.

1   a shoebox   _____

2   a roll of paper towels   _____

3   each of a pair of dice   _____

4   a Christmas tree   _____

5   a can of frozen orange juice   _____

6   a telephone book   _____

Use what you know about polygons and solids to answer the questions below.

7   What is the total area of all six sides of this cube?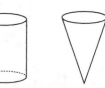

_____

8   Is this figure a rectangular solid?

_____

9   Which object would hold more water, this cylinder or this cone?

_____

Geometry Skills Practice

Circle the letter for the best answer to each problem.

1   Which parts of this figure are congruent?

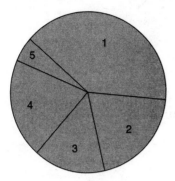

   A parts 1 and 5
   B parts 2 and 4
   C parts 2 and 5
   D parts 1 and 4

2   Which figures are similar to the figure in the box?

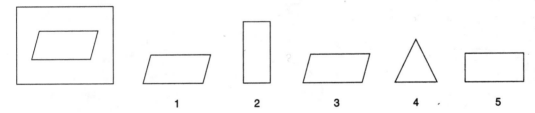

   F only figure 3
   G figures 1 and 3
   H figures 3 and 4
   J figures 1, 3, and 5

3   What is the measure of angle A?

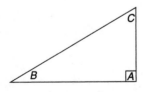

   A 30 degrees
   B 45 degrees
   C 90 degrees
   D 180 degrees

4 Which diagram shows parallel lines?

F

G

H

J

5 Which number is inside the square and the circle, but outside the triangle?

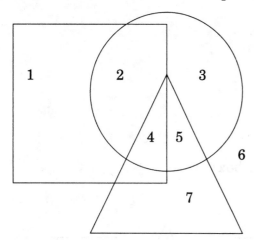

A 2

B 3

C 4

D 5

Skills Inventory Post-Test

Part A: Computation

Circle the letter for the correct answer to each problem.

1

$4 + 5 + 3 =$ _____

- A 13
- B 12
- C 9
- D 11
- E None of these

2

$461 + 38 =$ _____

- F 841
- G 741
- H 499
- J 489
- K None of these

3

$\begin{array}{r} 579 \\ - 174 \\ \hline \end{array}$

- A 415
- B 403
- C 475
- D 406
- E None of these

4

$852 + 9 =$ _____

- F 851
- G 861
- H 942
- J 1,752
- K None of these

5

$21 \times 34 =$ _____

- A 614
- B 147
- C 704
- D 714
- E None of these

6

$16 \times 10 =$ _____

- F 26
- G 116
- H 160
- J 106
- K None of these

7

$5\overline{)2055}$

- A 411
- B 4,011
- C 511
- D 1,511
- E None of these

8

$6,714 + 27 =$ _____

- F 6,984
- G 6,744
- H 6,731
- J 6,741
- K None of these

9

$6\overline{)30}$

- A 6   C 4
- B 5   D 24
- E None of these

10

$\begin{array}{r} 2,345 \\ - 165 \\ \hline \end{array}$

- F 2,180
- G 2,280
- H 2,170
- J 2,270
- K None of these

11

$\dfrac{40}{4} =$ _____

- A 10   C $\dfrac{1}{4}$
- B $\dfrac{1}{10}$ D 4
- E None of these

12

$\begin{array}{r} 302 \\ \times 3 \\ \hline \end{array}$

- F 605
- G 905
- H 65
- J 95
- K None of these

13

$\begin{array}{r} 302 \\ - 60 \\ \hline \end{array}$

- A 232
- B 242
- C 342
- D 332
- E None of these

14

$3,125 - 127 =$ _____

- F 3,002
- G 3,008
- H 2,998
- J 3,018
- K None of these

15

$113 \times 27 =$ _____

A 3,051
B 2,051
C 3,017
D 3,031
E None of these

16

$8 \div 5 =$ _____

F 1
G 2
H 1 r 3
J 1 r 2
K None of these

17

$0.3 \times 9 =$ _____

A 27
B 2.7
C 0.27
D 0.027
E None of these

18

$\dfrac{3}{5} + \dfrac{1}{5} =$ _____

F $\dfrac{2}{5}$
G $\dfrac{4}{5}$
H $\dfrac{4}{25}$
J $\dfrac{3}{5}$
K None of these

19

$\dfrac{3}{4} \times \dfrac{2}{6} \times \dfrac{3}{4} =$ _____

A $\dfrac{3}{16}$
B $\dfrac{1}{4}$
C $\dfrac{1}{8}$
D $\dfrac{3}{8}$
E None of these

20

$$\begin{array}{r} 0.2 \\ \times\ 0.4 \\ \hline \end{array}$$

F 8
G 0.8
H 0.08
J 0.008
K None of these

21

$$\begin{array}{r} \dfrac{4}{5} \\ -\ \dfrac{2}{5} \\ \hline \end{array}$$

A 2
B $\dfrac{2}{5}$
C $\dfrac{2}{10}$
D $\dfrac{2}{25}$
E None of these

22

$$\begin{array}{r} \dfrac{2}{6} \\ \times\ \dfrac{3}{4} \\ \hline \end{array}$$

F $\dfrac{1}{4}$
G $\dfrac{1}{3}$
H $1\dfrac{1}{2}$
J 4
K None of these

23

$$\begin{array}{r} 1,245 \\ \times\ \ \ \ 4 \\ \hline \end{array}$$

A 4,840
B 4,960
C 4,860
D 4,880
E None of these

24

$$\begin{array}{r} 12.3 \\ -\ 0.25 \\ \hline \end{array}$$

F 12.5
G 12.15
H 12.05
J 9.8
K None of these

25

$17 - 1.245 =$ _____

A 15.755
B 16.245
C 16.245
D 15.655
E None of these

Part B: Applied Mathematics

Circle the letter for the correct answer to each problem.

1 What number is missing from this number pattern?
43, 39, 35, 31, 27, 23, _?_, 15, 11

    A 22
    B 25
    C 19
    D 17

2 Which group of numbers is missing from this number pattern?
3, 6, 9, 12, _?_, _?_, _?_, 24, 27, 30

    F 14, 17, 20
    G 15, 18, 21
    H 15, 19, 22
    J 16, 18, 21

3 Which of these is the same as the number in the place-value chart?

thousands	hundreds	tens	ones
4	3	0	1

    A 431
    B four thousand, thirty-one hundred
    C 4000 + 300 + 10
    D four thousand, three hundred one

4 Which of these numbers is a common factor of 24 and 32?

    F 6
    G 3
    H 8
    J 12

5 If $5 \times n = 45$, then n is _?_ .

    A 8   **C** 9
    B 5   **D** 6

6 Erica has 45 pieces of costume jewelry. Each drawer of her jewelry box can hold 12 pieces. How many drawers can she completely fill?

    F 1   **H** 3
    G 2   **J** 4

7 Kendall works three days a week at a fast-food restaurant. Last week he served 65 people on Monday, 42 people on Tuesday, and 89 people on Thursday. To estimate how many people he served in all, what numbers should Kendall add?

    A 60, 40, and 80
    B 60, 40, and 90
    C 70, 40, and 90
    D 60, 40, and 100

8 What number goes in the boxes to make both number sentences true?
19 − 5 = ☐
19 − ☐ = 5

    F 12   **H** 13
    G 14   **J** 15

9 You earn an extra hour of vacation for every 5 hours of overtime you work. If you work 15 hours of overtime, how many hours of extra vacation time will you get?

    A 3   **C** 4
    B 5   **D** 10

This circle graph shows how Ramona spends her time on a typical weekday. Study the graph. Then do Numbers 10 through 13.

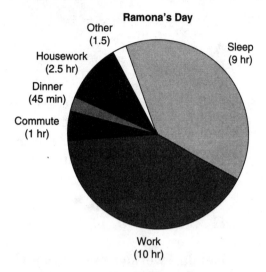

Ramona's Day

Other (1.5)
Sleep (9 hr)
Housework (2.5 hr)
Dinner (45 min)
Commute (1 hr)
Work (10 hr)

10 If Ramona cut in half the amount of time she spends on housework, how much more free time would she have each day?

F 2 hours   H $\frac{1}{2}$ hour

G 4 hours   J 1 hour

11 During a 5-day work week, how much time does Ramona spend commuting?

A 1 hour   C 5 hours
B 7 hours   D 10 hours

12 Ramona spends 4 times as much time working as she spends ? .

F sleeping
G commuting
H making dinner
J doing housework

13 Yesterday, Ramona spent two hours more at work than usual. How many hours did she work yesterday?

A 10 hours   C 6 hours
B 8 hours   D 12 hours

The information below shows how much a particular company charges for bus tours in New York City. Study the chart. Then do Numbers 14 through 17.

Tour Rates, per Person

1-hour tour of Broadway	$3.95
2-hour tour of SoHo	$9.00
3-hour tour of Midtown	$13.50

14 Which of these is most likely to be the cost of a 6-hour tour of the Manhattan?

F $14.00
G $17.00
H $27.00
J $30.00

15 If the cost of a lunch is *L,* then this number sentence shows the cost of taking a 3-hour tour with lunch.

$$\$13.50 + L = \$22.50$$

How much does lunch cost?

A $9.00
B $9.50
C $11.00
D $8.00

16 Which amount is closest to the exact cost to take 4 people on a 1-hour tour of Broadway?

F $9.00
G $16.50
H $17.00
J $16.00

17 Melissa takes a 1-hour tour of Broadway. She pays with a 20-dollar bill. How much change should she get back?

A $0.55
B $19.55
C $16.05
D $5.05

Amber is making a gingerbread house. This diagram shows the pieces of gingerbread she must make for the project. Study the diagram. Then do Numbers 18 through 24.

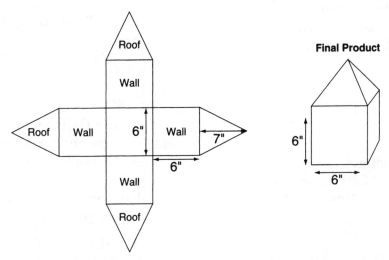

18 What is the area of each wall of the house?

    F 36 square inches
    G 12 square inches
    H 24 square inches
    J This cannot be determined.

19 What will be the perimeter of the bottom of the house?

    A 36 inches
    B 12 inches
    C 24 inches
    D This cannot be determined.

20 The walls of the house are all of the following except ___?___ .

    F quadrilaterals
    G squares
    H parallelograms
    J hexagons

21 Which of these cookie sheets could Amber use to bake each triangular roof section?

    A square sheet, 6 inches by 6 inches
    B a rectangular sheet, 4 inches by 7 inches
    C a square sheet, 7 inches by 7 inches
    D a circular sheet, 7 inches in diameter

22 When Amber puts this house together, she should form right angles at the corners where ___?___ .

    F one wall meets another wall
    G a roof section rests on a wall
    H the roof sections come together

23 To make this house, Amber must triple a recipe for gingerbread cookies. That recipe calls for $\frac{3}{4}$ cups of molasses. How much molasses should Amber use?

    A $\frac{9}{12}$ cup
    B $\frac{4}{9}$ cup
    C 4 cups
    D $2\frac{1}{4}$ cups

24 Amber must bake the dough for 12 minutes. She puts her first batch in at 3:54. When should she take it out?

    F 4:06
    G 4:04
    H 3:56
    J 3:42

25 Xavier wants to put a baseboard around the perimeter of his bedroom. The bedroom is 15 feet long and 10 feet wide. How much baseboard does Xavier need?

   A 50 feet
   B 25 feet
   C 150 feet
   D 5 feet

The chart below shows the sales tax charged in Central City. Study the table. Then do Numbers 26 through 28.

Sale	Tax	Sale	Tax
0–11 cents	1 cent	56–66 cents	6 cents
12–22 cents	2 cents	67–77 cents	7 cents
23–33 cents	3 cents	78–88 cents	8 cents
34–44 cents	4 cents	89–99 cents	9 cents
45–55 cents	5 cents		

26 The tax goes up one cent for every __?__ .

   F 9 cents spent
   G 11 cents spent
   H 12 cents spent
   J 13 cents spent

27 How much tax would there be on a $1.50 purchase?

   A 5 cents
   B 10 cents
   C 14 cents
   D 15 cents

28 How much would it cost to buy a piece of candy for 50 cents, plus tax?

   F 50 cents
   G 51 cents
   H 55 cents
   J 59 cents

Study this price list. Then do Numbers 29 through 33.

Tent Rental Fees

Tent Size	Rate per Day
10′ by 10′	$15.00
15′ by 10′	$22.00
20′ by 15′	$37.00

Pick up and Delivery: $25.00
Office Hours: 9:30 A.M. to 6:00 P.M., Monday through Friday

29 How much would it cost to rent two 10′ by 10′ tents for three days?

   A $15.00   **C** $60.00
   B $30.00   **D** $90.00

30 How much would it cost to rent a 15′ by 10′ tent for two days and have it delivered?

   F $44.00   **H** $47.00
   G $69.00   **J** $77.00

31 How much area would the 20′ by 15′ tent cover?

   A 300 ft²   **C** 70 ft²
   B 35 ft²   **D** 150 ft²

32 Maria is thinking about renting the 20′ by 15′ tent for her family reunion. If the cost of one-day rental were split among 9 families, about how much would each family pay?

   F $2.00   **H** $4.00
   G $3.00   **J** $5.00

33 How long is the office open each weekday?

   A $9\frac{1}{2}$ hr   **C** $15\frac{1}{2}$ hr

   B $8\frac{3}{4}$ hr   **D** $8\frac{1}{2}$ hr

The instructions below come from a bottle of cement color. Read the instructions. Then use them to do Numbers 34 through 37.

Mix $1\frac{1}{4}$ gallons of water with 10 ounces of color. Add this solution to 120 pounds of concrete mix. Mix thoroughly.

34 $1\frac{1}{4}$ gallons is between __?__ and __?__.

F $1\frac{1}{3}$ gallons and $1\frac{1}{2}$ gallons

G 1 gallon and $1\frac{1}{2}$ gallons

H $\frac{3}{4}$ gallon and 1 gallon

J $1\frac{1}{2}$ gallons and $1\frac{3}{4}$ gallons

35 $1\frac{1}{4}$ gallons is equal to how many cups?

A 16
B 18
C 20
D 5

36 Glenn wants to color 60 pounds of cement. How much color should he use?

F 4 ounces
G 6 ounces
H 8 ounces
J 5 ounces

37 The 10-ounce bottle of color costs $3.95. About how much does it cost per ounce?

A 30 cents
B 40 cents
C 50 cents
D 20 cents

38 Kevin was told he would earn three thousand, fifty-six dollars a month at his new job. Which of these numbers is three thousand, fifty-six dollars?

F $30,056
G $3,560
H $3,056
J $3,506

39 The decimal numbers below begin with 0 and a decimal point, then use the digits 0, 3, 5, and 7 exactly once. Which of the following are the smallest and the largest decimal numbers that can be written?

A 0.0357 and 0.7530
B 0.0537 and 0.7530
C 0.0357 and 0.7350
D 0.0753 and 0.7530

40 In the list below, "Input" numbers have been changed by a rule to get "Output" numbers. Which of these could be the rule for changing the "Input" numbers to the "Output" numbers?

Input	Output
3	4
6	5
9	6

F Add 1.
G Divide by 3. Then add 3.
H Multiply by 2. Then subtract 2.
J Add 1. Then subtract 4.

Members of the Blackwood Condominium Association all own apartments in the same building. They put their money together to pay for any building repairs. This graph shows how much the association had in savings each year from 1993 to 1998. Study the graph. Then use it to do Numbers 41 through 44.

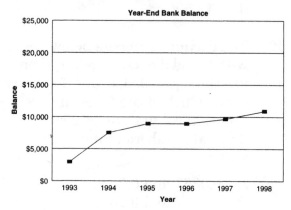

41 During which two years was the association's bank balance about the same?

    A 1994 and 1995
    B 1995 and 1996
    C 1996 and 1997
    D 1997 and 1998

42 Between what two years did the bank balance increase the most?

    F 1993 and 1994
    G 1994 and 1995
    H 1996 and 1997
    J 1997 and 1998

43 Which of these is the best estimate of how much the association had in its bank account at the end of 1998?

    A $10,000
    B $10,500
    C $11,000
    D $12,000

44 About how much money did the association save between 1993 and 1994?

    F $3,000
    G $8,000
    H $5,000
    J $2,000

45 If the same number is used in both boxes, which of these statements would be true?

    A If $\square - 6 = 8$, then $8 + 6 = \square$.
    B If $\square - 6 = 8$, then $8 - 6 = \square$.
    C If $\square - 6 = 8$, then $6 - \square = 8$.
    D If $\square - 6 = 8$, then $6 \times 8 = \square$.

46 Which number is inside the circle and the triangle, but outside the square?

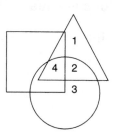

    F 1   **H** 3
    G 2   **J** 4

47 In the number 70,314, what is the value of the digit 7?

    A 70
    B 700
    C 7,000
    D 70,000

48 What will be the next figure in the pattern?

    F ≈   **H** ▦
    G ▦   **J** ▨

49 Which figure is congruent to the figure in the box?

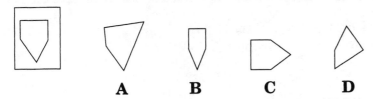

A   **B**   **C**   **D**

50 Which diagram shows perpendicular lines?

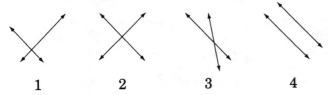

1   2   3   4

F 1 only
G 2 and 3 only
H 2 only
J 1 and 2 only

Post-Test Inventory Evaluation Charts

Use the key to check your answers on the Post-Test. The Evaluation Chart shows where you can turn in the book to find help with the problems you missed.

Keys

Part A

1	B	14	H
2	H	15	A
3	E	16	H
4	G	17	B
5	D	18	G
6	H	19	A
7	A	20	H
8	J	21	B
9	B	22	F
10	F	23	E
11	A	24	H
12	K	25	A
13	B		

Part B

1	C	26	G
2	G	27	D
3	D	28	H
4	H	29	D
5	C	30	G
6	H	31	A
7	C	32	H
8	G	33	D
9	A	34	G
10	J	35	C
11	C	36	J
12	J	37	B
13	D	38	H
14	H	39	A
15	A	40	G
16	J	41	B
17	C	42	F
18	F	43	C
19	C	44	H
20	J	45	A
21	C	46	G
22	F	47	D
23	D	48	F
24	F	49	C
25	A	50	J

Evaluation Charts

Part A

Problem Numbers	Skill Areas	Practice Pages
1, 2, 4,8	Addition of Whole Numbers	9–20
3, 10, 13,14	Subtraction of Whole Numbers	21–31
5, 6, 12, 15, 23	Multiplication of Whole Numbers	32–41
7, 9, 16	Division of Whole Numbers	42–55
17, 20, 24, 25	Decimals	56–64
11, 18, 19, 21, 22	Fractions, Ratios, Proportions	65–83

Part B

Problem Numbers	Skill Areas	Practice Pages
3, 4, 5, 8, 34, 38, 39, 45, 47	Numeration/ Number Theory	1–8, 9, 21, 32, 42, 56
10, 11, 12, 13, 27, 28, 41, 42, 43	Data Interpretation	84–96
1, 2, 14, 15, 25, 26, 40, 48	Pre-Algebra, Algebra	97–109
18, 19, 31, 33, 35	Measurement	110–126
20, 21, 22, 46, 49, 50	Geometry	127–140
6, 9, 23, 24, 29, 30, 36	Computation in Context	17–18, 28–29, 38–39, 52–53, 62, 77, 80–81
7, 16, 17, 32, 37, 44	Estimation, Rounding	5, 6, 16, 27, 37, 51, 76

Answer Key

Page 1,
Finding Place Value
1. hundreds, tens, ones
2. 1 thousand, 2 hundreds, 3 tens, and 0 ones
3. 3 ten thousands, 1 thousand, 4 hundreds, 0 tens, and 5 ones
4. hundreds
5. ten thousands
6. 3
7. 6
8. three hundred *or* 300
9. five thousand *or* 5,000
10. six hundred *or* 600

Page 2, Naming Large Numbers
1. three thousand, four hundred eighty-two
2. twenty-one thousand, six hundred one
3. four thousand, three
4. one thousand, forty-one
5. twenty-three thousand, ten
6. 902
7. 1,263
8. 52,000
9. 80,142
10. 4,306
11. 13,600
12. 21,585

Page 3, Comparing Whole Numbers
1. 2,842
2. 193
3. 21,403
4. 98
5. 876
6. 20,000
7. 51
8. 11,000
9. 160
10. 13
11. 52, 97, 115, 146
12. 9, 17, 95, 103

13. 178, 203, 517, 695, 961
14. 127
15. 678
16. 239

Page 4, Reading and Writing Dollars and Cents
1. three, twenty-three
2. four, three
3. one, thirteen
4. zero, fifty
5. ten, thirteen
6. fifty-one, ten
7. one hundred six
8. $8.59
9. $2.09
10. $10.50
11. $82.16
12. $0.25
13. $0.97
14. $0.08

Page 5, Estimating
1. A and D should be circled.
2. F
3. B
4. H
5. A

Page 6, Rounding
1. 80
2. 120
3. 100
4. $1.00
5. $5.10
6. 1,900
7. 80
8. 100
9. 20
10. $1.00
11. $1.00 *or* 1 dollar
12. 20
13. 90
14. 100
15. 2,120

16. 52,000
17. 870
18. 1,000
19. $2.00

Pages 7–8, Number System Skills Practice
1. C
2. H
3. D
4. G
5. A
6. G
7. A
8. F
9. D
10. H
11. D
12. J
13. B

Page 9, Basic Concepts
1. 3 + 6 = 9 and

$$\begin{array}{r} 3 \\ + \; 6 \\ \hline 9 \end{array}$$

2. C
3. 423
4. False

Page 10, Basic Addition Facts
1. 5, 6, 14, 3
2. 16, 10, 4, 15
3. 2, 11, 9, 4
4. 7, 11, 0, 5
5. 7, 15, 8, 17, 18, 4
6. 7, 12, 8, 6, 3, 3
7. 6, 8, 8, 12, 12, 10
8. 10, 11, 2, 9, 9, 5
9. 13, 10, 15, 7, 8
10. 13, 12, 16, 6, 13

Page 11, Simple Addition
1. 94; 381; 93; 817
2. 79; 6,553; 943; 95

3. 308; 939; 772; 8,460; 79; 77,998
4. 59 miles; 816 lb; 600; 88; 3,898
5. 80; 98 feet; 83 cents; 597; 71,000
6. $7.03; 776; $53.00; $93.00; 115
7. $4.24; 98 in.; 7,106; $6.82; $7.46;

Page 12, Adding Small and Large Numbers
1. 59; 288; $589.00; 1,510; 18 feet
2. 498; 8,063; 586 inches; 31,300; 596 gallons
3. 58
4. 76
5. $19.00
6. 4,685
7. 557
8. $816.00
9. 6,488
10. 199 min

Page 13, Adding Three or More Numbers
1. 63; 19; 10; 18¢; 39; 287
2. 167; 39 yards; 136; 11; 12; 15 in.
3. 57; 55
4. 170; 99
5. 109; 179
6. 166; 109

Pages 14–15, Adding a Column When the Sum Is More Than 9
1. 66; 118; 70
2. 190; 1,125; 88 cm
3. 271 gal; 11,502; 107 cans; 272 in.; 120
4. 225 miles; 6,528; 216; 919 in.; 818
5. 520; $40.50; $81.00; 240; 732; 323
6. $30.00
7. 228
8. $1.90

9. 651
10. 1,200
11. $125.00
12. $2.40
13. 3,110
14. 530
15. $8.25
16. 622
17. 452
18. 5,100
19. $68.12

Page 16, Using Estimation To Check Addition
1. 13,000; 11,000; 160 yards; $70.00
2. 800; 30,000; 5,000; 200
3. D
4. H
5. B
6. F
7. A
8. H

Pages 17–18, Solving Word Problems
1. B, D, E, G, and H should be circled.
2. 273
3. $8.90
4. $5.75
5. $8.00
6. how much the shirt cost
7. how much the cat weighed last year
8. the basic price of the car
9. how much he used to pay

Pages 19–20, Addition Skills Practice
1. D
2. G
3. A
4. J
5. E
6. J
7. C
8. G

9. B
10. J
11. A
12. G
13. E
14. H
15. B
16. J
17. D
18. J

Page 21, Basic Concepts
1. *F*
2. *T*
3. *T*
4. *T*
5. *T*

Page 22, Basic Subtraction Facts
1. 2, 4, 5
2. 2, 8, 3
3. 9, 1, 3
4. 3, 3, 1
5. 1, 1, 0, 1, 6, 4
6. 6, 2, 2, 2, 1, 3
7. 4, 4, 6, 4, 1, 7
8. 6, 5, 8, 7, 1, 5
9. 5, 0, 3, 5, 8
10. 3, 6, 2, 4, 4

Pages 23–24, Subtracting Larger Numbers
1. 250
2. 231
3. 1,410
4. 11 miles
5. 310
6. 202
7. 2,300
8. 30
9. $2.13
10. 120
11. $10.00
12. 54 in.
13. $2.27
14. 611
15. 721
16. 900
17. 1,200

18. 5,860
19. 812
20. 7.16
21. $8.70
22. $211
23. 6,014
24. 23.22
25. 31
26. 1
27. 31,112
28. 62,503
29. 51.38
30. 62.80
31. 112.00

Pages 25–26, Borrowing
1. 74
2. 127
3. 592
4. $0.72
5. 26
6. 1,493 meters
7. 1,037 feet
8. $11.25
9. $1.25
10. 863
11. 6,491 miles
12. 3,061
13. 683
14. 826
15. 424
16. $0.79
17. 992
18. 495 cm
19. 2,910
20. 9,949
21. 819
22. 3,975
23. 3,044
24. 239
25. 5.75
26. 6.75

Page 27, Using Estimation To Check Subtraction
1. $10.00
2. $300.00
3. 17,000 or 22,000
4. $30.00
5. 50,000

6. 2,000
7. 1,000
8. 60,000
9. $30.00
10. 7,000
11. 400 feet
12. $6.00
13. $37.00
14. $22.00
15. $4.00

Pages 28–29, Solving Word Problems
1. C
2. F
3. B
4. G
5. C
6. J
7. B
8.
$$\begin{array}{r} 892 \\ -\ 381 \\ \hline 511 \end{array}$$
9.
$$\begin{array}{r} 35 \\ -\ 15 \\ \hline 20 \end{array}$$
10.
$$\begin{array}{r} 892 \\ -\ 559 \\ \hline 333 \end{array}$$
11.
$$\begin{array}{r} 35 \\ +\ 15 \\ \hline 50 \end{array}$$
12. how much money he had to begin with
13. how much the dining room set costs
14. how much cement he uses for the patio
15. how long it usually takes Sylvia to get to work

Pages 30–31, Subtraction Skills Practice
1. D
2. F
3. A
4. H
5. D
6. F
7. D

8. J
9. C
10. J
11. A
12. G
13. B
14. F
15. D
16. K
17. D
18. F
19. B
20. F

Page 32, Basic Concepts
1. 4,562
2. 0
3. 65×4 or 4×65
4. $59 + 59 + 59$
5. 12×2 or 4×6
6. *T*
7. *F*
8. *T*
9. *F*
10. *F*
11. *T*
12. *T*

Page 33, Basic Multiplication Facts
1. 12, 16
2. 6, 32, 36
3. 30, 15, 42
4. 8, 28, 72
5. 12, 4, 21, 27
6. 49, 24, 9, 20
7. 45, 35, 14, 24
8. 16, 18, 40, 64
9. 36, 56, 25, 48
10. 63, 18, 81, 54

Page 34, Multiplying by a One-Digit Number
1. 48; 696; $4.88; 155; 246 inches
2. 408; 405 feet; $9.00; 148; 276
3. 428
4. 129
5. 350

6. 4,050
7. 99
8. $8.26
9. $18.88
10. 21,009
11. $64.00

Page 35, Multiplying by a Two-Digit Number
1. 143; 2,662; 3,926; 525; 903
2. 20,800; 9,150; 2,880; 1,953; 1,045
3. 6,930
4. 16,416
5. 2,130
6. 6,946
7. 968
8. 2,550
9. 19,800
10. 3,200
11. 9,460

Page 36, When a Column Product Is More Than 9
1. 160
2. 260
3. 1,530
4. 195
5. 224
6. 625
7. 1,232
8. 140
9. 570
10. 4,963
11. 1,250
12. 660
13. 8,568
14. 3,655
15. 817

Page 37, Using Estimation When You Multiply
1. Multiply 70 by 10; Estimate: 700
2. Multiply 100 by 30; Estimate: 3,000
3. Multiply 2,000 by 9; Estimate: 18,000

4. Multiply $30.00 by 3; Estimate: $90.00
5. Multiply $10.00 by 5; Estimate: $50.00
6. Multiply 90 feet by 20; Estimate: 1,800 feet
7. Multiply $100 by 5; Estimate: $500
8. Multiply 500 by 70; Estimate: 35,000
9. Multiply 80 by 80; Estimate: 6,400
10. Multiply 600 by 90; Estimate: 54,000
11. Multiply 800 by 60; Estimate: 48,000
12. Multiply 90 by 40; Estimate: 3,600

Pages 38–39, Solving Word Problems
1. C
2. H
3. H
4. A
5. $128.00
6. 26
7. 105
8. 400
9. what type of car he rented
10. how many people came to the wedding
11. how many square feet the patio will be

Pages 40–41, Multiplication Skills Practice
1. B
2. G
3. C
4. F
5. E
6. K
7. C
8. J
9. B
10. G
11. A
12. J

13. C
14. J
15. B
16. F
17. B
18. H
19. A
20. J

Page 42, Basic Concepts
1. A and D should be circled.
2. 671
3. 0
4. No
5. 1
6. 66

Page 43, Basic Division Facts
1. 2, 4, 2, 3
2. 2, 8, 6, 5
3. 6, 5, 3, 4
4. 5, 7, 2, 3
5. 4, 2, 7, 9
6. 7, 8, 7, 7
7. 9, 5, 2, 4
8. 8, 9, 8, 8
9. 9, 7, 5, 6
10. 7, 6, 9, 6
11. 6, 6, 3, 4
12. 8, 6, 9, 4

Page 44, Using the Division Bracket $\overline{)}$
1. $5\overline{)70}$
2. $2\overline{)98}$
3. $12\overline{)24}$
4. $12\overline{)142}$
5. $10\overline{)150}$
6. $25\overline{)575}$
7. $15\overline{)90}$
8. $9\overline{)81}$
9. $4\overline{)212}$ inches
10. $2\overline{)98}$ feet
11. $4\overline{)16}$
12. $5\overline{)\$20.00}$

13. $3\overline{)1200}$

14. $3\overline{)654}$

15. $12\overline{)250}$

16. $12\overline{)24}$

Page 45, Dividing by a One-Digit Number

1. 21
2. 32
3. 134
4. 112
5. 412
6. 101
7. 11 inches
8. 304 dollars
9. 321 inches
10. $4\overline{)844} = 211$
11. $5\overline{)5055} = 1011$
12. $2\overline{)26} = 13$
13. $4\overline{)84} = 21$

Page 46, When a Dividend Digit Is Too Small

1. 21
2. 21
3. 300
4. 31
5. 81
6. 420
7. 51
8. 31 inches
9. 61 pounds
10. 105
11. 109
12. 109
13. 104
14. 3,007
15. 406
16. 203
17. 209
18. 107 tons
19. 703
20. 109
21. 103

Page 47, Dividing with a Remainder

1. 1 r 1
2. 3 r 3
3. 4 r 2
4. 4 r 3
5. 2 r 1
6. 2 r 3
7. 2 r 1
8. 2 r 2
9. 5 r 1
10. 2 r 4
11. 2 r 2
12. 3 r 3
13. 10 r 1
14. 10; 2
15. 3; 3

Page 48, Writing the Steps of a Division Problem

1. 25
2. 29
3. 13
4. 27
5. 12
6. 14
7. 14
8. 121
9. 112
10. 115
11. 131
12. 620
13. 108
14. 181
15. 14
16. 24

Page 49, Long Division

1. 135
2. 22
3. 144
4. 22
5. 178
6. 165

Page 50, Dividing by a Two-Digit Number

1. 2
2. 3
3. 11
4. 31
5. 2
6. 3
7. 2
8. 22
9. 3

Page 51, Using Estimation When You Divide

1. 400
2. 60
3. 100
4. 100
5. 200
6. 100
7. 30
8. 20
9. 1,000
10. 20
11. C
12. G
13. B
14. H
15. C

Page 52, Solving Word Problems

1. 125
2. $12.00
3. 180
4. 223
5. $42.00
6. $1.00
7. $188.00
8. $360.00

Page 53, Solving Two–Step Word Problems

1. B
2. G
3. C

4. H
5. $90.00
6. $290.00
7. $25.00
8. $95.50

Pages 54–55, Division Skills Practice
1. B
2. J
3. C
4. J
5. C
6. F
7. A
8. H
9. D
10. F
11. A
12. H
13. C
14. G
15. B
16. G
17. D
18. H
19. B
20. J

Page 56, Decimals
1. hundredths
2. $\frac{1}{10}$ or one tenth
3. $\frac{3}{100}$ or three hundredths
4. hundredths
5. tenths
6. one tenth
7. 0.08
8. hundredths
9. thousandths
10. 6.3
11. 12.2

Page 57, Comparing Decimal Numbers
1. 0.003
2. 0.098
3. 0.00899
4. 1.032

5. 0.13
6. 0.005
7. 0.09
8. 0.19
9. 1.05
10. 0.002
11. 0.01, 0.1, 1, 11
12. 0.023, 0.032, 0.15, 0.75
13. 0.75, 2.3, 8.7
14. 0.0159
15. 0.0137

Page 58, Adding Decimals
1. 8.6; 0.46; $2.60; 1.14; 2.62; 1.03
2. 3.19; 0.19; 5.264
3. 0.198; 6.012; 1.82
4. 6.238; 52.95; 23.115

Page 59, Subtracting Decimals
1. 2.9; 0.03; $7.25; 0.05; 1.38, 7.48
2. 0.10 (or 0.1); 0.74; 0 .06
3. 6.47; 0.84; 11.187
4. 0.122; 0.743; 3.444

Pages 60–61, Multiplying Decimals
1. 0.18; 0.26; $27.00; $2.25
2. 0.75; 1.525; $49.95; $0.45
3. 13.2; 0.132; 59.52
4. 27.6
5. 6.4
6. 1.26
7. 1.56 million
8. 0.036; 0 .084; 2.10; 0.300; 0.042; 0.062
9. 0.0180; 0.0350; 0.0550; 0.0680; 0.0045; 0.014
10. 0.0525
11. 37.5 lb
12. 0.6 million
13. 1.5 feet

Page 62, Solving Word Problems
1. C
2. F

3. C
4. June
5. 4.5 feet
6. 421.3
7. 2.478 million
8. 29.2 million

Pages 63–64, Decimal Skills Practice
1. E
2. H
3. C
4. H
5. C
6. J
7. C
8. G
9. C
10. H
11. A
12. H
13. B
14. H
15. C
16. F
17. D
18. F
19. C
20. K
21. D
22. H

Page 65, Numerators and Denominators
1. $\frac{1}{2}$
2. $\frac{1}{4}$
3. $\frac{1}{6}$
4. $\frac{1}{8}$
5. $\frac{3}{4}$
6. $\frac{3}{8}$
7. $\frac{2}{5}$
8. $\frac{3}{8}$

9. The circle should be divided into 2 equal sections. One section should be shaded.
10. The circle should be divided into 8 equal sections. Five sections should be shaded.
11. The circle should be divided into 6 equal sections. One section should be shaded.
12. The circle should be divided into 2 equal sections. Both sections should be shaded.

Page 66, Fractions and Division

1. $\frac{27}{100}$
2. $\frac{153}{1760}$
3. $\frac{3}{52}$
4. $\frac{15}{23}$
5. $\frac{2}{9}$
6. $\frac{31}{175}$
7. $\frac{1}{15}$
8. $\frac{156}{500}$
9. 6
10. 2
11. 4
12. 5
13. 50
14. 10
15. 5
16. 5
17. 6

Page 67, Reducing a Fraction to Simplest Terms

1. $\frac{1}{2}$
2. $\frac{1}{5}$

3. $\frac{1}{5}$
4. $\frac{1}{5}$
5. $\frac{1}{3}$
6. $\frac{2}{5}$
7. $\frac{1}{2}$
8. $\frac{1}{2}$
9. $\frac{1}{5}$
10. $\frac{1}{4}$
11. $\frac{1}{3}$
12. $\frac{1}{4}$

Page 68, Finding Factors

1. 2, 5
2. 3
3. 2, 7
4. 3, 5
5. 2, 4, 8
6. 2, 4, 5, 10
7. 1, 5, and 25
8. 1, 2, 5, 10, 25, and 50
9. 1, 2, 3, 6, 9, and 18
10. 1, 2, 4, 5, 10, 20, 25, 50, and 100
11. 1, 2, 3, 4, 6, 8, 12, and 24
12. 12
13. 5
14. 25
15. 3
16. 4
17. 9
18. 4
19. 2
20. 10

Page 69, Fractions Equal To 1 and Fractions Greater Than 1

1. less than
2. equal to
3. less than
4. greater than
5. greater than

6. greater than
7. less than
8. less than
9. 2
10. 3
11. $1\frac{1}{8}$
12. $2\frac{2}{5}$
13. $1\frac{3}{5}$
14. 5
15. $3\frac{2}{5}$
16. $6\frac{1}{3}$
17. $1\frac{7}{10}$
18. $2\frac{1}{9}$
19. $5\frac{1}{5}$
20. $4\frac{3}{8}$

Page 70, Comparing Fractions

1. $\frac{3}{5}$
2. $\frac{2}{3}$
3. $\frac{1}{7}$
4. $\frac{2}{5}$
5. $\frac{1}{5}$
6. $\frac{1}{7}$
7. $\frac{2}{5}$
8. $\frac{3}{7}$
9. $\frac{4}{5}$
10. $\frac{1}{6}$
11. $\frac{3}{5}$
12. $3\frac{1}{4}$
13. $\frac{2}{5}, \frac{4}{5}, \frac{5}{5}, \frac{7}{5}$

14. $\frac{1}{5}, \frac{1}{4}, \frac{1}{3}, \frac{1}{2}$

15. $\frac{2}{9}, \frac{5}{9}, \frac{6}{9}, \frac{7}{9}$

16. $1\frac{1}{5}, 1\frac{1}{3}, 1\frac{1}{2}$

17. $\frac{3}{7}, \frac{3}{5}, \frac{3}{4}$

Page 71, Adding Fractions

1. $\frac{5}{6}$

2. $\frac{3}{7}$

3. $\frac{5}{9}$

4. $\frac{2}{5}$

5. $\frac{5}{8}$

6. $\frac{4}{9}$

7. $\frac{1}{3}$

8. $\frac{2}{3}$

9. $\frac{1}{5}$

10. $\frac{1}{2}$

11. $\frac{1}{4}$

12. $\frac{1}{3}$

13. $\frac{5}{5} = 1$

14. $\frac{8}{8} = 1$

15. $\frac{14}{7} = 2$

16. $\frac{14}{9} = 1\frac{5}{9}$

17. $\frac{7}{5} = 1\frac{2}{5}$

18. $\frac{7}{6} = 1\frac{1}{6}$

Page 72, Subtracting Fractions

1. $\frac{1}{6}$

2. $\frac{1}{5}$

3. 0

4. $\frac{1}{5}$

5. $\frac{1}{6}$

6. $\frac{1}{3}$

7. $\frac{1}{3}$

8. $\frac{3}{4}$

9. $\frac{1}{2}$

10. $\frac{1}{2}$

11. $\frac{1}{5}$

12. $\frac{1}{2}$

13. $\frac{2}{3}$

14. $\frac{3}{4}$

15. $\frac{1}{2}$

16. $\frac{1}{6}$

17. $\frac{1}{4}$

18. $\frac{4}{5}$

Page 73, Multiplying Fractions

1. $\frac{1}{6}$

2. $\frac{2}{5}$

3. $\frac{1}{5}$

4. $\frac{1}{24}$

5. $\frac{1}{9}$

6. $\frac{1}{6}$

7. $\frac{1}{9}$

8. $\frac{5}{21}$

9. $\frac{1}{7}$

10. $\frac{1}{4}$

11. $\frac{1}{12}$

12. $\frac{3}{49}$

13. $\frac{2}{25}$

14. $\frac{1}{8}$

15. $\frac{5}{81}$

16. $\frac{1}{6}$

17. $\frac{3}{16}$

18. $\frac{2}{15}$

19. $\frac{1}{12}$

20. $\frac{1}{20}$

21. $\frac{1}{8}$

22. $\frac{1}{10}$

Page 74, Canceling Before You Multiply

1. $\frac{2}{5}$

2. $\frac{1}{6}$

3. $\frac{3}{10}$

4. $\frac{1}{20}$

5. $\frac{2}{15}$

6. $\frac{1}{12}$

7. $\frac{1}{3}$

8. $\frac{2}{9}$

9. $\frac{3}{10}$

10. $\frac{1}{12}$

Page 75, Multiplying a Fraction by a Whole Number

1. 4
2. 2
3. $\frac{1}{2}$
4. 3
5. $\frac{2}{3}$
6. $1\frac{2}{5}$
7. $\frac{2}{3}$
8. $1\frac{1}{8}$
9. 3
10. $1\frac{1}{2}$
11. $1\frac{4}{5}$
12. 1
13. 5
14. $2\frac{2}{5}$

Page 76, Rounding a Fraction or a Mixed Number

1. 2
2. 3
3. 5
4. 6
5. 3
6. 1
7. 5
8. 9
9. 4
10. 1
11. 8
12. 8
13. 5
14. 5

Page 77, Solving Word Problems

1. too long
2. $\frac{1}{2}$
3. yes
4. $\frac{2}{3}$
5. $\frac{22}{125}$
6. $\frac{1}{2}$ cup
7. $\frac{12}{8}$ or $1\frac{1}{2}$ inches
8. 1 inch
9. $\frac{7}{12}$ foot
10. the baby
11. $\frac{3}{4}$ cup

Page 78, Comparing Numbers Using Ratios

1. $\dfrac{1 \text{ cup ginger ale}}{3 \text{ cups fruit juice}}$
2. $\dfrac{2 \text{ teachers}}{31 \text{ students}}$
3. $\dfrac{100 \text{ miles driving}}{15 \text{ minutes rest}}$
4. $\dfrac{4 \text{ quarts}}{1 \text{ gallon}}$
5. $\dfrac{22 \text{ miles}}{1 \text{ gallon}}$
6. $\dfrac{3 \text{ cans}}{1 \text{ dollar}}$
7. $\dfrac{\$12.00}{1 \text{ family}}$
8. $\dfrac{1 \text{ manager}}{4 \text{ employees}}$
9. $\dfrac{2 \text{ lost}}{1 \text{ won}}$
10. $\dfrac{3 \text{ hours babysitting}}{5 \text{ hours driving}}$
11. $\dfrac{4 \text{ soda cans}}{1 \text{ juice can}}$

Page 79, Finding a Unit Rate

1. 20 miles per gallon
2. $5.00 per ounce

3. $0.80 *or* 80 cents per candy bar
4. $75.00 per performance
5. 18 miles per hour
6. $12.00 per hour
7. 400 miles per day
8. $1.80 per pound

Pages 80–81, Writing Proportions

1. $\dfrac{2 \text{ hours}}{25 \text{ cars}} = \dfrac{8 \text{ hours}}{100 \text{ cars}}$
2. $\dfrac{60 \text{ words}}{1 \text{ minute}} = \dfrac{180 \text{ words}}{3 \text{ minutes}}$
3. $\dfrac{12 \text{ eggs}}{\$1.00} = \dfrac{6 \text{ eggs}}{\$0.50}$
4. $\dfrac{1 \text{ inch}}{50 \text{ miles}} = \dfrac{2 \text{ inches}}{? \text{ miles}}$
5. $\dfrac{2 \text{ days}}{50 \text{ pots}} = \dfrac{20 \text{ days}}{? \text{ pots}}$
6. 2
7. 30
8. 6
9. 9
10. 100
11. 24
12. 6
13. 6
14. $7.80
15. 15
16. 20 sets
17. 16 cans
18. 32 children
19. $10.50

Pages 82–83, Fractions Skills Practice

1. D
2. J
3. B
4. K
5. B
6. J
7. D
8. A
9. H
10. D
11. G
12. B
13. F

14. B
15. J
16. C
17. J
18. C
19. H

Pages 84–85,
Reading a Table
 1. D
 2. G
 3. C
 4. F
 5. $2.00
 6. $6.50
 7. $5.25
 8. a regular video for one day
 9. $7.50
10. $5.00
11. new releases
12. computer software
13. C should be circled

Page 86, Using Numbers in a Table
 1. Linda Hansen
 2. David Blume
 3. 5 years
 4. 5
 5. 4
 6. Tubman
 7. Roosevelt
 8. 2

Page 87,
Using a Price List
 1. $1.60
 2. $1.41
 3. $3.99
 4. $2.50
 5. $17.40
 6. $2.50
 7. 10
 8. 3 cents

Page 88,
Median and Mode
 1. 3 minutes
 2. 23

 3. 65
 4. 2
 5. 23
 6. 78

Page 89, Finding an Average
 1. 40 years
 2. 20 minutes
 3. 41 minutes
 4. A: $2\frac{1}{2}$, 2, 3; B: 4, 3 *or* 5 , 4; C: 2, 2, 2; D: 6, 5 *or* 6, 6

Page 90, Graphs
 1. 5
 2. 8
 3. 6

Page 91,
Reading a Circle Graph
 1. B
 2. J
 3. D
 4. F
 5. B

Page 92,
Reading a Bar Graph
 1. 4,000
 2. H
 3. D
 4. workers
 5. 2
 6. $\frac{4}{11}$
 7. A

Page 93,
Reading a Line Graph
 1. 5
 2. 10
 3. Accept any answer between 16 and 18.
 4. 5 minutes
 5. 4
 6. the 10th and the 11th

 7. B
 8. G

Page 94, Using a Graph
 1. 25
 2. 5
 3. 20
 4. 30

Pages 95–96, Data Interpretation Skills Practice
 1. B
 2. J
 3. C
 4. G
 5. D
 6. H
 7. B
 8. F
 9. D
10. F
11. B

Pages 97–98, Patterns
 1. C
 2. grey
 3. C
 4. up
 5. 5 and 6
 6. two dots, bar, X (*Wording will vary.*)
 7. A
 8.
 9.
10.

   4 parts   10 parts
11.

Answer Key

Page 99, Finding Number Patterns

1. 7, 4, 1
2. 21, 27, 33
3. 18, 23, 28
4. 16, 32, 64
5. 15, 31, 63
6. fours
7. 11
8. 3
9. 3
10. a number plus 7
11. a number plus 15
12. a number minus 5
13. a number minus 6
14. 18
15. 35, 20
16. 11, 13
17. 8
18. 16
19. 7

Page 100, Patterns in Number Sentences

1. $+$
2. $-$
3. $\times$
4. $+$
5. $+$
6. $-$
7. $\div$
8. $\div$
9. $+$
10. $-$
11. 4
12. 8
13. 10
14. 12
15. 4
16. 9
17. 12
18. 22
19. 2
20. 10
21. 12

Page 101, Some Basic Number Properties

1. $\div$
2. $-$
3. $\times$ or $\div$

4. -84
5. -488
6. $\div 12$
7. T
8. T

Page 102, Functions

1. Add 5.
2. Subtract 3.
3. Add 11.
4. Multiply by 2.
5. Add 4.
6. Subtract 8.
7. Add 12.
8. A
9. H

Pages 103–104, Writing Letters and Symbols for Words

1. $n + 32$
2. $n \times 3$ or $3n$
3. $n + 24$
4. $n + 6$
5. $n - 7$
6. $\frac{1}{2} \times n$ or $\frac{n}{2}$ or $n \div 2$
7. $n + 8$
8. $n \times 6$ or $6n$
9. $13 \div n$ or $\frac{13}{n}$
10. $n - 6$
11. $14 \times n$ or $14n$
12. $n + 10$
13. $n \times 2$ or $2n$
14. $n - 5$
15. $\frac{3}{4} \times n$
16. $2 \times n$ or $2n$
17. $9 - n$
18. $n/12$
19. $n \times 10$ or $10n$
20. $r + 20$ dollars
21. $p \div 4$ or $\frac{p}{4}$
22. $x - 450$ dollars
23. $3 \times c$ cents or $3c$ cents
24. $d + m$
25. $4 \times k$ minutes or $4k$ minutes
26. $28 - d$ students

27. $6 - F$ cups
28. $y - 7$ years
29. $172 - x$ pounds

Page 105, Writing Equations

1. Sample: $7 + x = 12$
2. Sample: $5 - 1\frac{1}{2} = x$
3. Sample: $7 + 3 = x$
4. Sample: $65 + 15 = x$
5. Sample: $x + 6 = 15$
6. Sample: $24 - 7 = x$
7. Sample: $12{,}000 - 9{,}000 = x$

Pages 106–107, Solving Equations

1. 8
2. 25
3. 15
4. 25
5. 39
6. 40
7. 15
8. 13
9. 103
10. 6
11. $x + 7 = 15$; 8
12. $x - 6 = 7$; 13
13. $20 \div 4 = x$; $5.00
14. $35 + 12 = x$; 47
15. $28 = 2 \times x$; 14
16. $6.25 = $4.50 + $1.25 + x$; $0.50
17. $x = 8 + 2 \times 8$; 24

Pages 108–109, Algebra Skills Practice

1. D
2. G
3. A
4. F
5. C
6. F
7. B
8. G
9. C
10. F
11. C
12. F

13. D
14. G
15. C

Page 110, Choosing the Best Tool
1. length
2. weight and/or height (length)
3. capacity
4. capacity
5. length
6. length
7. time
8. time
9. temperature
10. weight
11. B
12. E
13. D
14. E
15. G
16. C
17. F
18. A

Page 111: Reading a Scale
1. $1\frac{3}{4}$ in.
2. $2\frac{1}{4}$ in.
3. $3\frac{1}{2}$ in. or $3\frac{2}{4}$ in.
4. $2\frac{1}{4}$ in.
5. 3 in.
6. $2\frac{3}{4}$ in.
7. $7\frac{1}{4}$ in.
8. 5 in.
9. $4\frac{1}{2}$ in. or $4\frac{2}{4}$ in.
10. 1 in.

Pages 112–113, Reading a Scale that Skips Numbers
1. 13
2. 45
3. 10
4. 5

5. 15
6. 65
7. 150
8. 95
9. 2
10. 2
11. 4
12. 5
13. 10
14. 5, 40
15. 2, 64

Page 114, Measuring Temperature
1. degrees Fahrenheit
2. 32°F
3. colder
4. D
5. one degree Celsius
6. 2, 2
7. 96°F–98°F; 36°C–38°C
8. B

Page 115, Measuring Length
1. C
2. F
3. B
4. H
5. A
6. 24
7. 3
8. 6
9. 2
10. 3
11. 18
12. 15
13. 3

Pages 116–117, Using a Ruler
1. $2\frac{1}{2}$ in.
2. $1\frac{1}{2}$
3. 1 in.
4. $\frac{1}{2}$ in.
5. 2 in.
6. $\frac{3}{4}$ in.

7. $1\frac{1}{4}$ in.
8. 1 in.
9. $\frac{1}{4}$ in.
10. $1\frac{1}{2}$ in.
11. 12 in. by 4 in.
12. $3\frac{1}{2}$ in. by 2 in.
13. $10\frac{1}{2}$ in. by 8 in.
14. 8 in. by $10\frac{1}{2}$ in.
15. $9\frac{1}{2}$ in. by 4 in.
16. $4\frac{5}{8}$ in. by $1\frac{1}{4}$ in.
17. $12\frac{1}{2}$ in. by $12\frac{1}{2}$ in.
18. $8\frac{1}{2}$ in. by 11 in.
19. $1\frac{5}{8}$ in.
20. $\frac{3}{8}$ in.
21. $\frac{5}{8}$ in.
22. $1\frac{1}{8}$ in.
23. $2\frac{2}{16}$ in. or $2\frac{1}{8}$ in.
24. $6\frac{3}{16}$ in.
25. $\frac{15}{16}$ in.
26. $\frac{15}{16}$ in.
27. $\frac{11}{16}$ in.

Page 118, Metric Units of Length
1. 12
2. 5
3. $2\frac{1}{2}$
4. 9
5. 90
6. 10
7. 70
8. 500
9. 0.3

Page 119,
Finding Perimeter
1. 34 feet
2. 30 mi
3. 30
4. 10
5. 8 cm
6. 22'
7. 16 feet
8. 2 cm

Page 120, Finding Area
1. 36 centimeters^2
2. 8 mm^2
3. 50 mi^2
4. 36 in.^{2}
5. 4 ft^2
6. 4 cm^2
7. 18 m^2
8. 5 inches

Page 121,
Measuring Weight
1. 8
2. 24
3. 500
4. 250
5. B
6. J
7. 19
8. 2.55
9. 250
10. 3 grams
11. 900 grams
12. 1 pound

Page 122, Measuring
Liquid Capacity
1. A
2. G
3. D
4. H
5. a gallon
6. 1 gallon
7. 6 liters
8. 6
9. 12
10. 8
11. 2
12. 24

13. 4
14. $1\frac{1}{4}$

Page 123,
Measuring Time
1. 30
2. 180
3. 2
4. 14
5. 48
6. 15
7. 20
8. 90
9. 12
10. 5 hours, 45 minutes *or* $5\frac{3}{4}$ hours
11. 1 hour, 30 minutes *or* $1\frac{1}{2}$ hours
12. 6 hours, 20 minutes *or* $6\frac{1}{3}$ hours
13. 9 hours
14. 5 hours, 24 minutes

Page 124,
Changes in Time
1. 9:00
2. 11:00
3. 4:20
4. 4 hours, 20 minutes *or* $4\frac{1}{3}$ hours
5. 1:45

Pages 125–126,
Measurement Skills
Practice
1. D
2. H
3. B
4. H
5. A
6. H
7. A
8. F
9. B
10. J

11. A
12. G
13. C
14. H
15. C
16. H
17. C

Page 127, Using Logic
1. C
2. 4
3. C
4. A
5. B
6. 75

Page 128, Points, Lines,
and Angles
1. *A*
2. *B*
3. *C*
4. angle
5. *A, B, C, D,* and *E*
6. 2
7. *B, A,* and *E* or *D, A,* and *C* (in any order.)

Page 129, Angles
1. A
2. A
3. A
4. B
5. circled
6. circled
7. not circled
8. circled
9. not circled
10. not circled
11. 6 and 7
12. 8
13. 1, 2, 3, 4, and 5

Page 130, Lines
(*The answers to each question may be given in any order.*)
1. B, E, F, G, H
2. A, C, and D
3. F and H

4. Oak Street, Avenue B, and Avenue A
5. Avenue B and Avenue A

Page 131, Polygons
(*Answers may be arranged in any order.*)
1. A, D, K, N
2. B, C, F, G, I, L
3. E and M
4. J
5. F
6. E, F, J
7. B, F, G, I, L
8. regular, parallelogram, quadrilateral

Page 132, Triangles
1. 45 degrees
2. 30 degrees
3. 50 degrees
4. 110 degrees
5. 48 degrees
6. 30 degrees
7. 60 degrees
8. 1 and 5
9. 7
10. 1, 2, and 7

Page 133, Circles
1. A and C
2. B
3. 4 in.

4. 6 cm
5. cannot tell
6. 50 mm
7. cannot tell
8. $1\frac{1}{2}$ cm

Page 134, Drawing Figures
1. B
2. F
3. C
4. F
5. A

Page 135, Symmetry
1. D
2. F
3. D
4. H
5. D

Page 136, Congruent Figures
1. C
2. G
3. A
4. B
5. F
6. A and B (or C and F)

Page 137, Similar Figures
1. B
2. G
3. B
4. yes
5. no
6. C
7. None

Page 138, Three-Dimensional Figures
1. a rectangular solid
2. a cylinder
3. a cube
4. a cone
5. a cylinder
6. a rectangular solid
7. 24 square inches
8. no (Its corners do not form right angles.)
9. the cylinder

Pages 139–140, Geometry Skills Practice
1. B
2. G
3. C
4. H
5. A

Glossary of Common Terms

angle: the figure formed by two lines coming from the same point.

average: one way of describing the most typical number in a set of numbers. The average (or mean) is the sum of all the values in a set divided by the total number of values. Example: To find the average of the set 2, 4, 6, and 8, add all four numbers together. Then, divide the sum (20) by 4. The average is 5.

canceling: a way to simplify multiplication problems that contain fractions. To cancel, use the same number to divide one of the numerators and one of the denominators in the fractions being multiplied. Example:

$$\frac{2}{5} \times \frac{3}{4} = \frac{\overset{1}{2} \times 3}{5 \times \underset{2}{4}} = \frac{3}{10}$$

carrying: used in addition and multiplication problems when the sum or product of a column has two digits. The tens digit is added to the sum or the product of the next column. Examples:

$$\begin{array}{r} {}^{1}3\overset{1}{4}7 \\ + \ 285 \\ \hline 632 \end{array} \qquad \begin{array}{r} 1\overset{4}{0}9 \\ \times \quad 5 \\ \hline 545 \end{array}$$

congruent figures: figures with the same size and shape

decimal: a number containing digits to the right of a small dot called the **decimal point.** These digits represent part of one. Example: In $13.42, digits to the left of the decimal point represent 13 whole dollars, while digits to the right represent 42 hundredths of a dollar.

diameter: the width of a circle

digit: 0, 1, 2, 3, 4, 5, 6, 7, 8, or 9. The number 145 has three digits, 1, 4, and 5.

dividend: the number in a division problem that is being divided. Example: In the problem 30 ÷ 6 = 5, thirty is the dividend.

divisor: the number in a division problem that you are dividing by. Example: In the problem 20 ÷ 4 = 5, four is the divisor.

equation: a number sentence showing that numbers or mathematical expressions are equal. Example: $x + 7 = 8$

equivalent fractions: fractions that have the same value. Examples: $\frac{1}{2}$, $\frac{4}{8}$, and $\frac{5}{10}$ are all equivalent fractions.

estimate: a number that is close to the actual or exact amount

factor: a number that can evenly divide a larger number. Examples: 1, 2, 4, 5, 10, and 20 are all factors of 20.

fraction: a way of representing part of something. The bottom number in a fraction (the **denominator**) tells how many parts there are in the whole. The top number (the **numerator**) shows how many parts the fraction stands for. Example: If there are 23 people in your whole class and 5 of those people are missing, $\frac{5}{23}$ of the class is absent.

intersecting lines: lines that cross or that will cross

inverse operations: addition and subtraction or multiplication and division. Inverse operations undo each other. Example: If 12 has been added to a number, you can find the original number by subtracting 12.

lowest terms: a fraction is in lowest (or simplest) terms when there is no whole number, other than 1, that can evenly divide both the top and the bottom numbers in it.

mean: *See* average.

metric units: the units of length, capacity, and weight used in Canada and many other parts of the world. Examples: centimeter, liter, gram

mixed number: a combination of a whole number and a fraction. Mixed numbers represent a value between two whole numbers. Example: $2\frac{1}{2}$

parallel lines: two lines that are always the same distance apart

perimeter: the total length of the sides in a figure

perpendicular lines: two lines that cross (or that will cross) to form right angles

place value: the location of a digit in a number. Example: 573 has three place values. The place value of the digit 5 is hundreds and of the digit 7 is tens.

point: a dot or a single location in space

polygon: a flat shape or figure that has straight sides. Examples: rectangle, hexagon, pentagon.

product: the result when you multiply

proportion: an equation showing that one ratio equals another. Examples:

$$\frac{3}{4} = \frac{6}{8} \text{ or } \frac{4 \text{ ounces}}{13 \text{ cents}} = \frac{16 \text{ ounces}}{52 \text{ cents}}$$

quotient: the result when you divide

radius: the distance from a point on a circle to the center of the circle

ratio: a comparison of two numbers. Example: Ron has 3 days off for every 7 days he works. Ratios can be written in three ways: 3 to 7, 3 : 7, or $\frac{3}{7}$.

reduce a fraction: divide both the top and bottom numbers in a fraction by the same number. This does not change the value of the fraction. Example: You can reduce the fraction $\frac{10}{20}$ by dividing both the top and the bottom by 10. $\frac{10}{20} = \frac{1}{2}$

remainder: the "left over" portion in a division problem

right angle: a 90-degree angle. Each corner in a square is a right angle.

rounding: increasing or decreasing a number so that it ends in one or more zeros. Example: 189, rounded to the nearest 100, is 200. Rounded numbers are close to the exact amount, but they are easier to work with.

similar figures: figures with the same shape, but not necessarily the same size

simplest terms: *See* lowest terms.

standard units: the units of length, capacity, and weight used in the United States. Examples: inch, pound, cup

symmetry: a figure is symmetric if it can be divided into two identical halves. The line that separates the two identical parts of such a figure is called the **line of symmetry.**

unit rate: the rate for one unit of a given quantity. Examples: 35 miles per (1) gallon or 3 cents per (1) ounce

unknown: a letter (or variable) in an algebraic expression. Example: In the expression $2a + 3 = 11$, a is the unknown.

variable: *See* unknown.

Index